Harry Saddler is a Melbourne-based writer. His writing about the interactions between people, animals and the environment has been published in *The Lifted Brow*, *Meanjin* and *The Guardian*, among others. His book *The Eastern Curlew* was shortlisted for the Queensland Literary Awards in 2019.

A CLEAR FLOWING YARRA

HARRY SADDLER

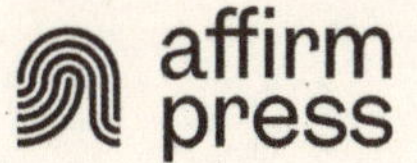

For Dad

You showed me the joy of nature

You made me curious

I'll always miss you and I'll always love you

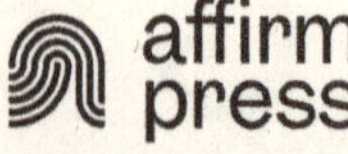

First published by Affirm Press in 2023
Boon Wurrung Country
28 Thistlethwaite Street
South Melbourne VIC 3205
affirmpress.com.au

10 9 8 7 6 5 4 3 2 1

A catalogue record for this book is available from the National Library of Australia

ISBN: 9781922848116 (paperback)

Cover design by Steph Bishop-Hall © Affirm Press
Typeset in Garamond Premier Pro by J&M Typesetting
Proudly printed in Australia by McPherson's Printing Group

AUTHOR'S NOTE

I'd like to thank everyone at Affirm Press but particularly Laura Franks and Martin Hughes for their patience, understanding and consideration – the creation of this book went through more than its fair share of horseshoe bends that nobody foresaw. Thank you to Mum, Dad and my brother Owen for your support and encouragement – not just recently, but for as long as I've been around. Thank you Emily for – just everything, really. Thank you to everyone who agreed to be interviewed, and who entrusted me with their ideas and words – I hope when you read the book it's something that you're pleased to be a part of.

This book is not intended to be a comprehensive encyclopedia of every single thing to do with the Birrarung/ Yarra. It's deliberately idiosyncratic and personal. It's a selection of visions: of the river as it is and the river as it could be. The hoary old saying goes that you can't step in the same river twice – but I think it's just as true to say that nobody's river is the same as anyone else's. This is my river – I hope you enjoy it, and I hope it encourages you to find your river, too.

CONTENTS

PROLOGUE

In 2017, on a hot March day, I went looking for azure kingfishers. Tiny, elusive and even bluer than their name suggests, they'd been seen several times on the Yarra River in the outer north-eastern suburbs of Melbourne. I took my bike on the train and rode to the river from Lilydale Station along the busy roads that pulled like taut wire between the paddocks and empty lots out there, nearly crashing at the bottom of a fast, steep hill while cars and trucks whooshed past me. I made it to the river eventually but I never found the kingfishers, and unwilling to go back the way I'd come I decided instead to ride back along the river vaguely in the direction of home. I wasn't dressed for a long ride: as happens more often than I'd care to admit I'd decided that jeans were an appropriate form of clothing for this activity, and as always I was wrong. There was no direct path along the river that I could find, or at least not one that was suitable to ride along on a bike, so I found myself detouring until I was almost lost, relying more and more on the map on my phone. I stumbled upon creek paths and followed them as far as I could, eager to avoid roads; because I'd been riding as much as possible along the river valley, when inevitably I did hit a road it took me up a steep hill. It was mid-afternoon and the sun stayed high and hot for hours, barely seeming to move. I grew more and more tired, got more and more disoriented in a maze of unfamiliar roads and paths. It probably took me no more than an hour of riding but it felt like much longer by the time I finally made it back into something that seemed like the Melbourne I knew: Warrandyte, still far from any train station and even further from home but at least familiar in its layout, streets and shops and houses and parkland. And as if in a dream, I found there a remarkable sight: hundreds of people, swimming and bathing in a wide, sparkling stream. I wasn't dressed for swimming but I dismounted and took off my helmet and, scooping up handfuls of fresh, clean water I splashed them over my head, washing the sweat away and cooling myself down. It felt beautiful.

Melbourne is a city of many creeks and rivers. I assumed that this clear stream, full of people, in a suburb I'd never been to before, must be some larger creek or secondary river that I wasn't familiar with. So when I eventually got home and looked up where I'd been, I was astonished by what the map told me: I'd washed my face in the Yarra River.

~

I grew up swimming in fresh water. The nearest sea beach was a two-hour drive from my home, but Canberra's Lake Burley Griffin was just a few blocks away. Swimming in the lake was a thing that Canberrans did then, in the 1980s and 1990s. Or else swimming in the Cotter River, or the Molonglo River, or the Murrumbidgee, amid water-curved rocks and coarse sand and casuarinas. I never learned how to spot a rip in the ocean but I learned not to dive to the bottom of rivers lest I get caught in a snag, not to swim in a river after heavy rain, that water in a river is heavier and faster than it looks and that a fast-flowing river can carry you away and drag you under before you even know what's happening. I learned the feel of algae and mud between my toes before the river bottom dropped away; I learned how to walk across submerged rocks, gritting my teeth against stubbed toes as the water rose coldly over my knees, over my belly, over my chest, and my pale skin disappeared into the dark tanniny stream. I learned about the shock or welcome relief of pockets of cold water beneath the surface.

I've never been much of a swimmer but the swimming I did always seemed at odds with what I was told, again and again, was the Australian way: golden, sandy beaches and saltwater spray and crashing waves. To this day I can't quite get my head around the appeal of swimming in sea water only to then have a shower on the beach to wash the salt off. If you've ever swum in a healthy river, or a healthy lake, then you know the feeling of coming out cleaner than you were when you went in. There's no other swimming like it.

But not every river is swimmable. When I was very young I had a book of jokes. I can barely remember any of them (which is probably for the best) but I have a dim recollection of some – including one, more a rhyme than a joke, in which the teller recounts a wish to be a sparrow, living free of care or obligation. Except sparrow, in the joke, is spelled 'sparrer', in order to force a rhyme: the punchline is that in this imagined, idyllic existence, the hours would be passed defecating in the Yarra.

That was my introduction to the largest river in Australia's second-largest city: that it was fit only to be the butt of jokes. And it was an attitude that was reinforced for me when I moved to Melbourne as a young adult where, like generations of new Melburnians before and since, I settled in the inner north, near where the Yarra flows. When I arrived I found the attitude shared by everyone who lived here – particularly those born and bred in Melbourne. Newcomers to Melbourne learn to view the Yarra as a dividing line between north and south, a seemingly impenetrable barrier in the city's consciousness – but not so much as an entity, and a collection of ecosystems, in its own right.

Sure, people use the Yarra: they row on it, they walk along it. But as a whole Melburnians don't embrace their river (and I should acknowledge here that it's only one of Melbourne's rivers: for many residents of this sprawling city 'the river' is the Maribyrnong, or the Werribee, or the Plenty). If you read about the Yarra it won't be long before you hear it called, derisively, 'the upside-down river', because of its cloudy, clay-filled water; if you mention that you've enjoyed spending some time along its banks it won't be long before some wag asks if you saw any bodies floating by. And these detractors have a point: until recently, in the history of the city of Melbourne, the Yarra's been treated as nothing better than a drain, and often as something considerably worse. Along its length you'll have no trouble finding rubbish floating on its surface or caught in the vegetation on its banks, and rubbish traps choked with all the refuse of a city of five million people. Any pleasant view of the Yarra is only a bend in the river away from being ruined by a pile of floating garbage, and the river's banks are havens for environmental weeds of all kinds.

But it's a lot better than it used to be. Consider this passage from Kristin Otto's *Yarra: A Diverting History*:

> There were slaughterhouses or abattoirs, along with every conceivable sidebar business, up to Dights Falls: fellmongers to prepare the animals' fresh, bloody skins for tanneries to make leather on the open ground; wool scourers to wash the fleece (in the 'fresh' water on the riverbanks); starch and glue factories and bone mills, to process hooves and horns; candle and soap factories. And all the unusable waste – heads, legs, megalitres of gross liquid – emptied into the flow ... Water taken from the Pumps at one point was described as having the consistency of very weak gelatine. In its lower reaches, the Yarra was a vomit-inducing sewer.

It's not just the absence of today's ubiquitous plastic waste that makes the horror of this passage unrecognisable from the Yarra that we know today. The hoary old saying tells us that we can't step in the same river twice, and it's true that the Yarra has constantly changed – and more to the point, *been* changed – throughout Melbourne's history. But so deplorable has been Melbourne's treatment of its main river that the city seems, collectively, to have forgotten that things can be changed for the better as well as for the worse. The Yarra has been abused for so long that as a community we've forgotten how to treat it any other way.

Although I've been in Melbourne now for nearly half my life, I still struggle to think of myself as a Melburnian: I don't follow the footy; I don't know where half the suburbs are; I don't understand most of the shibboleths. I cling to Canberra, my home town. But that changes when I'm by the Yarra: that old, slow river makes me feel more connected to Melbourne than anything else ever has. When I first started falling in love with the Yarra I got my wires crossed, and I thought that its muddy colour was natural – but I subsequently read that the muddiness of the river as we know it now is a result of erosion along its banks and the banks of its tributaries, and that that erosion was in turn a result of widespread land clearing since European colonisation. Before this process all began, so it's said, the Yarra ran clear. Before this process all began, so it's said, the Yarra ran clear. It's a jarring thought, and it should shake us out of how we regard the Yarra that we know now. And it should prompt us to ask ourselves, and the Melburnians around us: if the Yarra once ran clear from its source to its mouth, what would it be like to live once again along a clear flowing river?

A MESSY RIVER IS A HEALTHY RIVER

It was on May Day that I met the Yarra Riverkeeper. It was cold and grey, and damp hung in the air. The river was swollen with recent rain, the bush in disarray after unsettling storms. I caught the train to Eltham, where nature reaches into and among the houses and rare species can appear unexpectedly and vanish again just as suddenly. It was a Sunday morning, and our rendezvous had already been once postponed: I'd misinterpreted the messages the Riverkeeper had sent to me, and missed my window of opportunity. I resolved that I'd not make the same mistake again. The train's horn echoed in the stillness as I disembarked onto the nearly empty platform and removed my mask. She was already waiting, and welcomed me as though we were old friends, before ushering me into her vehicle.

'In terms of a job title,' I ventured, 'it's almost Tolkienesque, isn't it?'

Charlotte Sterrett laughed. 'You tell kids you're the Yarra Riverkeeper and they automatically envision what that is, and that's kind of how you want it,' she told me. 'And then you talk to an adult and they're like "But what do you do? What does it mean?" and it's lots of things, it's so varied.'

I asked her how she became the Yarra Riverkeeper.

'Well I just applied for the job!'

She'd been the Yarra Riverkeeper for only a few months, since January. She wasn't the first: as we drove past her street she pointed out a house. 'So this house here, this is Ian Penrose's house. It backs onto the river. And he was the first Riverkeeper. So I always knew about the Riverkeepers because of him.' After Ian Penrose was Andrew Kelly, and then Charlotte. As we drove through her neighbourhood she explained in more detail how she decided to apply for the position: 'We have a Warrandyte climate action group, we've got the local Osborne Peninsula Landcare, so I was already active in my personal life, but I've always wanted to do more for the Birrarung. After three or six years – they're three-year appointments – I'll have learned so much more about the river and be better able to advocate for her health and protection. I'm a social scientist but I'm not a scientist. So there's also something in it for me as well. I love to learn.'

The Yarra Riverkeeper Association has been around 'since 2004, 2005', Sterrett told me. 'We're fully independent ... we're primarily an advocacy group, and we advocate for the health, protection and care of the river and all her inhabitants from source to sea, including all her tributaries. We're also stewards. This includes advocacy on policy and legislation, as well as education, citizen science, regeneration and pollution. Advocacy is because all the best intentions in the world won't make the river healthy.'

It sounded like a lot of work, so I asked her how many people there were in the organisation, and Charlotte replied: 'We're small but nimble. We're very focused on the power of people and active citizens. So, we're all advocates, including our volunteers. Everyone who works here has a deep relationship with the Yarra, Birrarung. We speak from the heart and from on-the-ground experience. We do this work in service of the Birrarung, her traditional custodians – the Wurundjeri Woi-wurrung and Bunurong Peoples – and all Victorians.'

We'd left the car behind by the time Charlotte explained all this to me. She wanted to lead me on a walk: a twisting, winding route through flood debris and storm-tossed trees, along familiar paths and searching for routes now buried. It was a walk into history as well as towards the river. 'So [this] was all old farmland,' Charlotte told me as we set off, 'and you'll see some of the farm when we walk around the bend.' As we walked beneath tall, slender eucalypts she said: 'A lot of this, the trees aren't that old except the ones right along the river, eighty years old. There's obviously a lot of tea-tree, it's probably the main one, and it is considered a bit of a weed because it can take over quite a bit, but I don't mind it too much.' She led the way. 'Because if I'm going first I'm going to be the one who gets all the spiders in the face.'

She took me to one of her favourite parts of the river. 'We come swimming in here' – she and her family, she meant. Swimming in the Yarra didn't necessarily come naturally, though. Charlotte grew up in England, and at first she was unsure about immersing herself in the murky waters of the Yarra. 'I got over it pretty quickly. I always loved coming to Warrandyte and going swimming, and we used to just swim where the bridge is, which is where most people go, but there's plenty of other places to swim if you know about them. There used to be two log seats along here, and finally after all these years they're gone. That flood last year [in 2022] finally did it in. I used to bring my daughter here all the time and we used to set up a tarp ... and just spend the day sitting by the river. Lovely. She used to love using that as a play area for her little fairies and stuff ... we've seen wombats, platypus and kingfishers. I'm definitely a river person.'

As we walked along the river we stepped carefully: Charlotte pointed out trees overhead that had had their crowns snapped off in a recent storm; underfoot the ground was steep and sometimes slippery. 'It can be a bit muddy getting in and out after the rain,' Charlotte said of her swimming hole, and we got talking about mud, and clay, and the tiny clay particles that become readily suspended in the water and lend it its brown colour. I asked her about something I'd read on Wikipedia, that the river was clear before Europeans arrived – it's stated as fact, but with no citation to back it up, no evidence. 'I've wondered about that myself,' Charlotte told me, 'so I asked around and while no-one knows for certain, it's very likely that the river was clear all along its length. Urban development, habitat destruction; anything that disturbs the land causes soil to run off into the river making it the brown colour she is so famous for.'

She told me about the river as she'd seen it over the years: 'I take photos of the same place over and over again because every time you go it's just "Oh goodness it's so beautiful", and then you're like "I've taken enough photos" ... but each time the photo's different and sometimes the river looks really, really orange, and other times it looks more green, and sometimes it looks quite clear, so it does vary, it goes up and down.' But perhaps we fixate too much on how the water looks. 'Just because it looks muddy doesn't mean it's necessarily dirty, but it does tell you something is happening. Nutrients from sewage, chemicals, pesticides and stormwater are largely invisible and a massive threat to the Birrarung's health,' Charlotte pointed out as we walked. 'There are many other factors at play, but we use that one because it's a visual one that we can see. I also think that there's something about this European view of what nature should look like, and people in Australia tend to be brought up still to this day in that view that "this is beautiful, this isn't beautiful", and so I think Australian rivers or rivers like the Birrarung get a bad rap because they don't look as people expect them to look.' I recalled a lesson I'd been taught as an undergraduate biology student: sometimes a clear river can be a river in distress, because that clarity comes from an absence of life.

It's not just the colour of the river that changes. Charlotte pointed out a tree that had fallen in the river. 'Lots of people comment on trees that have fallen in the river after storms as if this is wrong, but actually a messy river is a healthy river,' she explained. 'Five years ago that tree was fully standing up with the best rope swing, and because this is an eddy it's really quite deep and we used to swim in there, and then one day it just fell over and it's moved more downstream every time [the river's] flooding. This has happened since the river was formed.'

And the riverbank changes, too. We only walked for an hour, maybe less, but Charlotte was at pains to point out, 'Even on this little walk that we're doing, the vegetation changes a lot. So [here] there's a lot more ferns, and this part here's a lot more wet, and you'll notice when we get to the next section it gets really dry, because it's facing north and the ground is rocky with little soil. You may not be able to notice because it has been done so well, but a lot of the plants you see are from reveg ... the local Landcare group has had a huge positive impact. Like many volunteer groups that care for the river, they have an ageing population, which is a concern.' Charlotte explained that the local Landcare group on the Osborne Peninsula where we were walking was just one of many small, hyper-local groups the Yarra Riverkeeper Association works with, 'and one thing we're finding is they've got an ageing volunteer workforce, and it's hard to recruit people when so many people have to work full-time to provide for their families.'

Revegetation is an ongoing effort; late in our walk Charlotte told me, 'We're coming into deer destruction territory. This had all been revegetated some years back. Because this was the original property on Osborne Road … So you can see a lot of European trees around here, and willows used to be everywhere as well. There's only one willow left. And when we first moved here this area was just terrible, so many weeds, but it's much improved. That is, until the deer came a few years ago.' She indicated the area we were walking through. 'It was a classic kind of degraded area here, then they planted it all out and had all bushes and trees and grasses, but now there's only really the grasses left, the deer eat everything.'

I asked Charlotte where the main focus of the Yarra Riverkeeper Association was: just on the river itself, or also on its tributaries? 'There's about fifty named tributaries, and then hundreds of creeks, but we can't advocate for all of them due to our small size. The catchment is very large – 4,000 square kilometres, in fact,' she replied. 'A third of Victoria's population lives in the catchment and we get 70 per cent of Melbourne's drinking water from the river, which has come at great cost to her health. She is also home to a third of Victoria's animal species. So that's why we have the network with the other local groups who care for their part of the river.' After our walk,

Charlotte introduced me via email to an almost overwhelming number of people and organisations, all working on their part of the river or its tributaries. On our walk, she went on to explain: 'Our historical focus has very much been on the main stem of the Birrarung, but we're looking beyond that now. There's lots of talk about the catchment as a whole and the concept of the Great Birrarung Parkland and what it might look like.'

She told me that the Association was starting to look at some of the key confluences, and had recently partnered with the University of Melbourne focusing on some of the creeks that flow into the river. 'So we're talking about ... how can we restore these areas and have net gain environmentally and culturally, and increase our understanding of what the history was, what the future could look like where the river can be healthy, protected and loved ... Each Riverkeeper comes in with their own ideas and I'm still formulating mine.' At this early stage of her tenure in the role, at least, her focus was higher upstream than had been the case for the two former Riverkeepers: 'There's a lot of agricultural land, golf courses, private land, there's flood plains and billabongs that used to be properly connected and functioning next to the river, so what would it look like to see these areas restored to full ecological health? But those are really tricky issues because it's about land tenure and many years of land practices that harm the river. And a culture of seeing the river as a resource rather than as a living entity with rights of her own. So this type of work takes a long time, and you've got to have the right relationships with people, and trust. Usually people go for the quicker wins, but I would like for us to look at that.'

I told Charlotte that the Yarra seemed, to me, like a gentle river – at least when it wasn't in flood. She agreed at first, but then hesitated: 'The Birrarung has many attributes. Around Warrandyte she's quick and has rapids, but in other parts she's slow and meandering. She can be loud and sometimes quiet, but never silent. The summer river is very different to the winter river. I've discovered parts of the river that I never knew, and every day she surprises me. I don't know, maybe people who don't spend so much time on the river don't notice that.'

Charlotte spoke about the river and about Australian landscapes with the enthusiasm of a convert. 'And actually that's why I did [an] outdoor education degree in the early 2000s,' she told me, 'because I moved to Australia and I felt really alienated from the natural environment because it was so different from what I'd grown up with, everything just felt wrong in a way. I had to re-educate myself. So I spent a year spending time in different environments, mostly in Victoria, learning about rivers, rock, land, sea, all the different environments – mountains, snow ... and that immersion in the environment changes your DNA a bit.'

We finished our walk, and got back in her car so she could drive me back to the train station. In the car we talked about public perceptions of the river: about attitudes towards it that are stuck in the past, of how so many people will tell you that it's a filthy, dead river. 'I really want to change that,' Charlotte said. 'It's so important. I have felt a bit of blow-back from a couple of people when I've been interviewed, and they're like "You're not talking about the issues enough! You're not talking about all the things that are wrong!" and I feel like people won't get interested in issues if it's all doom and gloom. I've worked in climate change for twenty years and I've learned that if you tell people the world's ending they just give up. So you have to have a balance, people have to feel that there's hope.' Which is not to say the river is pristine: it's still polluted. 'And I definitely want to talk about that,' Charlotte agreed. We drove towards Eltham, the river sliding past somewhere down the hill, out of sight, and Charlotte finished her thought: 'But I also want people to understand that they can't write her off as a lost cause, because she's not a lost cause. The Birrarung is beautiful and resilient even in the face of climate change, pollution, and urban development. She's worth fighting for.'

PLATYPUS

We had no chance of seeing them, of course. They were impossible to see: you had to get up at dawn or squint through the gloom of dusk. You had to sit long hours on a riverbank, shivering in the dewy grass. You couldn't just rock up to a suburban park with the sound of traffic a few blocks away and people walking dogs and riding bikes in the middle of a weekend afternoon, like you were breezing in for a picnic, and expect to see a platypus. A platypus! It was absurd, a laughable notion to think that we'd even be in with a chance on this cold and sunny day. That's what my friend and I agreed on as we walked along the path, admiring the anachronistic cows in their paddocks with their attendant cattle egrets, pointing out the kangaroos to each other. All this, and in the suburbs of Melbourne! But not platypuses too, no way. Still, we said, we're here now so we may as well try. We won't see one, of course, you never actually *see* one. But we may as well at least try. After all, people say ...

It took us about thirty seconds. My friend and I lost our shit. We gawped and we gaped at each other in delighted amazement and we laughed in mind-bent wonderment as a dark-brown platypus bobbed up to the surface of the pale-brown Yarra and then dived again, disappearing instantly into the turbid water. And then up again, and then down. Every few seconds: it never stayed down for long but it never stayed up for long, either. We didn't care. We were too busy grabbing each other's arms, pointing and staring, grinning in delight. As iconic and incredible an animal as exists anywhere in the world and it was here, in the Yarra, in the *Yarra*. And we'd walked here from a bus stop!

In hindsight that was my real awakening to the wildlife wonders of the urban Yarra, and as awakenings go it's hard to beat. While we were staring in joyous disbelief at the platypus, another person there that day, a local and an old hand at this kind of thing, told us that just a couple of kilometres further upstream, in a large suburban bush park bordered by the river, if we stood and watched a certain tree in the carpark at dusk, we'd see sugar gliders. That started a whole other obsession, one which continues to this day, and I'll tell you all about that later. Or two obsessions, really: an obsession with the wildlife of the Yarra River and its surrounds – but also an obsession just with the river. That maligned, too-often-misunderstood river, a river that we think of as urban and whose urbanness defines so much of our thinking about it, but that is just as much a rural river, a small-town river, even in some secret places a wild river. And it's a beautiful river, though many Melburnians will tell you otherwise or will just plain laugh at you if you call it that. I've seen it sparkling, that muddy old waterway, on a sunny day in the city where it's broad and slow and the only chop on its surface comes from the wake of boats: I swear it sparkles in the sun like a million diamonds. I've seen it providing home to all manner of wildlife: I've seen powerful owls fly across it at dusk; I've seen a white goshawk glare at me across the river's width from the branches of a river red gum, the back half of a ringtail possum in its lethal talons; I've seen grey-headed flying foxes dive-bomb into the river after dark on a hot January night, dipping their chests into the water, scaring the bejesus out of any duck unfortunate enough to be bobbing about at the wrong place and the wrong time, and then flapping laboriously up to the branches of the nearest tree to hang there upside down and lap the thirst-quenching water from their fur. I've seen the river rise and fall, flash-flooding to swallow everything or baring the gums of its muddy banks in the hunger of drought. I've even swum in it (and I'll tell you all about that later, too). Every time I open a tap and fill my bottle in Melbourne I'm drinking from it. I haven't been the first to experience any of these things, not by a long shot, and I won't be the last, but they've all shaped how I feel about the river, and from the river they've shaped how I feel about Melbourne, and from Melbourne they've shaped how I feel about my place in Melbourne, coming up now to twenty years living in this city. If a river is a city's lifeblood then the Yarra is the lifeblood of my connection to Melbourne: I probably wouldn't have left Melbourne if there were no river, but I certainly wouldn't have enjoyed my time here half as much. The Yarra's always given Melbourne a lot like that, and Melbourne's always taken a lot from the Yarra and never really bothered to ask if that's what the river might want. What could a river want, anyway? Well maybe what a river wants are the collective wants of everything that lives in and along and thanks to it.

So anyway, returning to those platypuses. I go back all the time now, of course. And I tell people about it like I'm an expert even though I'm really just an enthusiast, which anyway is a pretty bloody good thing to be, I think. I lived through the 90s so I've done my dash with disaffected faux-slackerdom: let's be passionate for a change! Let's be earnest and enthusiastic instead. Doesn't that sound like a whole lot more fun? So let me tell you about the time I saw eight platypuses at once in that same stretch of the Yarra. It's true, I swear. Of course they weren't all up at the same time, bobbing up and down like synchronised swimmers so you could count them off on your fingers, but all the same if a platypus goes down below on one side of the river and then straight away a platypus rises to the surface all the way over on the other side of the river, it's a pretty safe bet that those are two separate platypuses. Then factor in variations like size, or the fact that one big bruiser, the most massive platypus I've ever seen, had a big red cut under one eye and so couldn't be mistaken for any other individual – extrapolate from there, and you've got eight. Probably. Anyway that's my story and I'm sticking to it.

Let me tell you how that same year, that same September breeding season, I saw two platypuses swimming nose-to-tail on the surface in what I guess was some kind of courtship ritual.

Or how when I go to that spot I always see a platypus above a line of rapids, but lately I'm starting to wonder if that's just confirmation bias because I'm always looking at that part of the river because that's where I tell myself I always see a platypus. Either way I'm pretty sure there's a burrow in the bank near there, somewhere hidden under the long overhanging grass.

Or how I read on someone's blog that platypuses like cold,

grey weather, and someone else who was also there looking for them one day told me that they're very rarely seen in the warmer months, and how both of those things seem borne out in my half-decade or so of occasional observations.

Or how one time, when the Yarra was in raging flood and all the rapids had disappeared because the river was so high that it was *all* flood, a friend and I – different friend – saw a platypus swim up a side creek which was flowing more gently, and it seemed a lot like that platypus was tired and looking for a break.

Or how after years of going to that same spot – still the only place I've ever seen platypuses in the wild – I learned that the rocks in the river there which form the rapids that the platypuses hunt in are actually an old Indigenous fish trap, specially constructed.

Or how, just, bloody hell: *there are platypuses in the Yarra!* In the suburbs of Melbourne! A city of five million people and there are just platypuses swimming around down there and you can go see them any time you want, even in the middle of the day! (Though maybe not in January.) One time I was miles away, in the Dandenong Ranges National Park, and I saw two echidnas, and I finished my walk and I got on the train to take me back into the city, and then I thought: *hell, why not* – so in the city I hopped on a bus and forty minutes later I was watching two platypuses swimming around in the river and I thought: *how about that, I've seen two monotreme species in one day*. Which is a completely arbitrary and basically meaningless achievement that I call the Monotreme Double and it makes me unreasonably happy to know that I did it (actually, technically it was a double Double: two individuals of each species). Once when I was waiting for a platypus to show itself I even saw, in the distance way up high and circling above the river, a wedge-tailed eagle. Can you believe that someone once wrote, not so long ago in a serious national newspaper, 'Melbourne isn't exactly known for its wildlife'?

Let me tell you what I've learned about platypuses, and I'm going to assume that the ones I've seen in the Yarra are representative of the whole even though I've never seen them anywhere else except in captivity in Healesville Sanctuary, upstream in the Yarra Valley. They always face upstream when they come to the surface. They don't stay on the surface for long: just long enough to catch their breath, and then they're back under (actually I've read that they only ever eat on the surface, but I haven't looked closely enough to notice that). They only paddle with their front feet; their back feet drag behind them in the river's flow. They can ride rapids, surfing on or just beneath the surface down chutes of water over slippery rocks to get to a still pool beneath. They don't look like a log or a rock or anything else that you might have heard they look like: they look exactly like a platypus, and there's nothing else that you could mistake one for.

Let me tell you also, and you might already know this, that they're one of the very few venomous mammal species in the world: the males have venomous spurs behind their feet. Though the venom won't kill a person I've read that the pain from it is so severe that you might wish you were dead, and it can last for weeks or even months, and it can't be treated with opiates. Like that platypus going over the rapids, you've just gotta ride it out. (A quick bit of useful trivia, and skip this if you know it already too: what's the difference between venomous, poisonous and toxic? Toxic can hurt you, but that's not its primary purpose; it's just an unfortunate side effect. Poisonous is meant to hurt you, but only if you ingest it. Venomous hurts you too, but with intent: venom's on a sting or in a bite. Thanks, undergrad biology lessons.)

The first thing we all learn about platypuses in school is that Europeans back in Europe thought they were a hoax. If they'd known then what we know now about platypuses they would've thought they were tripping. Platypuses are carnivorous – so far, so normal – but the way they find their prey is something else: their bills are lined with hundreds and hundreds of sensors that can detect the electrical signals from the nerves of their prey as it moves around. Imagine you're a yabby; imagine your human legs are yabby legs. Stretch your leg out, or lift your arm (your yabby arm) from your side. Zap! There goes a little bolt of electricity. Now the platypus is coming to find you. (By the way, a few million years ago there was a giant platypus, a metre long, that had teeth to chew through turtle shells and lobster carapaces.) We're always told that platypuses are primitive mammals: they've been around for so long, and that whole egg-laying thing is a bit weird, a bit of a throwback – but so-called modern mammals haven't come up with anything half as cool as electrical sensors to find a yabby with a twitchy leg under a rock on the bottom of a muddy river.

Because that's another thing that always amazes me when I see the platypuses in the Yarra: once they dive they're *gone*. The water's that murky that they disappear from sight instantly. It's part of what makes seeing one in the Yarra so exciting: you never know when or where or if one's going to turn up because the window of opportunity in which you might get to see one before it goes below the surface again is so brief. So they're obviously not getting around under there by eyesight.

But they're there, and they seem to be doing pretty well for themselves. At least in that one little bit of the river, anyway.

In the Yarra. In the beautiful Yarra. Melbourne, how lucky are we!

A GATHERING OF EVERYTHING

When I moved to Melbourne in 2004 I did what countless numbers of new arrivals have done before me: I lived in Fitzroy. All the people I knew from Canberra who'd moved down here lived in the inner north so I followed them. Since then I've moved a handful of times but I've always stayed nearby – always, coincidentally, within the inner-city council known as the City of Yarra. With nearly one hundred thousand residents and a population density of four and a half thousand people per square kilometre, it's not somewhere you'd necessarily expect to have a lot of wildlife – let alone a biodiversity team.

'We're a team of three now,' Craig Lupton told me in his office one afternoon in early December 2022. Craig is the council's senior biodiversity officer. Stuck to the computer in his office was a photo of a powerful owl; an enormous, barred feather from a powerful owl also decorated his workspace. Indeed I first met Craig when I was watching a powerful owl in my local park and he literally stepped out of the bushes, having been monitoring the same bird. 'No successful breeding the past couple of years,' he told me when we met again in his office, 'but we know a few years back there was a record of two chicks.'

As the name indicates, the City of Yarra is dominated by the river: the entire western and southern borders of the council follow the meandering route of the Yarra as it works its way towards the CBD. 'We basically manage [the lower Yarra] from Alphington, so just a bit upstream from where Darebin Creek comes in, right down to Swan Street,' Craig told me. 'So we've got several waterways, Darebin Creek and Merri Creek and the Yarra River.' He explains that this means not only is there a north–south corridor for wildlife, but also an east–west link 'along Merri Creek parklands through to Royal Park and the like'.

Craig's been working with biodiversity for 'nearly thirty years, so I did twenty-odd years with Parks Victoria in a range of different metro parks, and now council, but it's great to see the work that was done along the Yarra ... and see the wildlife starting to come back'. He explained: 'We're getting black wallaby regularly through here, so Burnley right through Alphington – they're not common but you see them. We had [a] reporting of an echidna just opposite Coulson and Rushall [reserves]on the Merri Creek just recently, reportings of sugar gliders down at Alphington ... And we're seeing movement of species like the powerful owl, which is now resident here.'

Speaking at length to Craig, it soon became apparent that his knowledge of the wildlife in the City of Yarra was matched only by his enthusiasm. Throughout our conversation he kept thinking of species he hadn't yet mentioned: 'We're seeing new animals turn up all the time,' he said. For example: 'Peron's tree frog, which is not a usual species for this location, but that's quite stable now down at Alphington Wetlands – if you go down on an evening now in spring/summer you'll hear them going off their heads.' And where there's one species, others will follow: 'Tawny frogmouths, we see a lot of them through here and they do eat frogs, but it builds up the whole ecosystem doesn't it? It brings in other species, provides the food that they need, and then it's about the habitat and the structure.'

He's worked in the area for long enough to have seen it change significantly. When I mentioned to him a photo I'd seen of the land around Merri Creek in the 1980s, when it was nothing but grass and willows, compared to today when it's lush bushland, he gushed: 'The transformation there – [thanks to] Friends of Merri Creek, Merri Creek Management Committee – has been astronomical, it's been really quite amazing. And that's becoming a functional [wildlife] corridor too.'

The changes haven't just been in vegetation, but also in approaches to reinstating that vegetation: 'Best practice at the moment ... is around trying to create EVCs [Ecological Vegetation Classes],' Craig explained. EVCs are defined by the Victorian Department of Energy, Environment and Climate Action as 'the standard unit for classifying vegetation types in Victoria'.

> EVCs are described through a combination of floristics, lifeforms and ecological characteristics, and through an inferred fidelity to particular environmental attributes. Each EVC includes a collection of floristic communities (i.e. lower level in the classification) that occur across a biogeographic range, and although differing in species, have similar habitat and ecological processes operating.

'Years ago when I first started it wasn't about EVCs,' Craig told me, 'it was about planting tubestock of local indigenous species as thickly as you can with everything you can possibly put into a site, and then letting it establish and letting it do its own thing, and it might mean going in and intervening and cutting down the trees you put in because you put far too many trees in ... Back then that was the way we did it, now it's more ... looking at lifeform benchmarks and percentages of strata, x number of canopy trees per hectare, so you're doing it a lot more scientifically. More ground covers, shrubs and less overstorey. So it's changed, the way land managers and ecological practitioners are working. So in my time – it's only been thirty years – I've noticed that fundamental change.' Among other things, this new approach is more open to the idea of dynamic, ongoing ecological systems, which means that there's no real end to the work. 'None!' Craig agrees. 'That's why we have our bushland contracts [with contracted bushland-regeneration companies] go nine years ... which is unusual, but a long-term contract to achieve ecological change isn't in done one or two years, it takes time.' It's a timeframe that looks beyond – 'a long way beyond' – the standard term of a city council.

But it's not just time that transcends human delineations: an ecosystem doesn't end at the council boundary. 'A challenge urban land managers face,' Craig told me, 'is consistency of management on both sides of river and tributaries. Sometimes rivers are managed by different state and local government organisations with help from traditional owners and amazing support from community groups and residents. Funding to improve ecological values can vary a fair bit from side to side, because of different priorities. Good communication and sharing of resources between organisations can help to manage both sides of a river or creek.'

Craig mentioned that the Yarra Strategic Plan (Burndap Birrarung burndap umarkoo) 2022–32 is 'bridging the gap, so that's bringing councils and other state land management departments together and looking at not just their own patches and silos, it's about how do we influence the whole river and the whole catchment and how can we work with traditional owners to make it happen'. He described it as a 'massive paradigm shift', and explained: 'It's taking us a while to get our head around how that works and how we need to look beyond boundaries, because wildlife doesn't acknowledge boundaries ... we're near the [mouth of the river] down here [in the City of Yarra] so we're at the end of the line for everything that happens up there [upstream]. The Yarra Strategic Plan is already assisting in collaborative decision-making by involving traditional owners as decision makers on Country, alongside state and local government organisations and offering funding to ensure the Birrarung is being managed and protected as one living and integrated natural entity.'

He summarised the council's approach: 'We're targeting

sustainability with our vegetation, so trying to get it to a point where we don't have to intervene too much.' That means getting to a point where local vegetation can outcompete weeds but, he added, 'With disturbed landscapes it's always going to be difficult to have minimal intervention, it's going to get to the state where you'll maybe get natural regeneration or recruitment of one species over another, then you may have to intervene. But [we're] trying to get to a point where it does its own thing and creates its own habitat niches and opportunities.' Planting is in some ways more important than pulling out, and it's not as simple as just 'local good, introduced bad': 'Quite often you'll see that the riverbanks are left with a plethora of weeds that's holding them together, it's really important because we need the bank stabilisation and structure.' With flash floods becoming more severe and more frequent, Craig said, 'If you don't have bank structure or integrity what does that actually mean for the overall system?' Instead the council focuses on the land further back from the river: 'We try to be a bit realistic about where we work and how we work; it's very labour intensive to be working right in the riparian zone because ... you're going to keep getting floods and in the event of more severe weather events this is going to keep happening and it's going to happen worse probably, so we try to acknowledge that we're not [working] on the very edges.'

And, as Craig was quick to agree, a river is about more than just what happens in the water or at the water's edge: 'It's the association of vegetation with that wetness and moisture and its own environment that's so critical ... water feeds life, it creates life, it creates niches, it creates a gathering – it's a gathering of everything. It's so critical. And that's where our biodiversity hotspots are.'

Much of that biodiversity has been lost locally, but in addition to the species that have found their way back via newly re-established greenbelt connections, Craig envisages the possibility of more deliberate reintroductions. 'That's potentially an option out at Alphington Wetlands,' he told me, 'because it doesn't often connect with the Yarra, [so] it should be relatively free of introduced fish. And we've done a bit of work with some aquatic scientists on fish species within Alphington Wetlands, but more recently we've seen carp in there; I think with the floods [at the end of 2022] we've got carp in. But that's potentially a reintroduction area for threatened fish species, so that's something that's really interesting. It's hard to know, there's so many different factors about sites re-establishing native wildlife.'

As far as protecting landscape more generally, Craig and

the council are keen to create refuge areas for wildlife, but that can be a problem in an inner-urban area such as the City of Yarra. 'Dogs are a big problem for us, not only defecation but also harassing of wildlife and also trampling,' Craig told me, 'but you put up a fence in Yarra [and] the community can get upset, they don't like having any of their movements restricted. They like being able to go where they want to, when they want to.' The pandemic and lockdowns haven't helped: 'We know fencing's one of the most effective techniques for regeneration of bushland in areas where you've got higher traffic,' Craig said – but that traffic can be obstinate. 'We had at least eleven or twelve BMX areas throughout three or four bushland sites. We ended up putting a fence through Burnley Park through the bushland there because during COVID-19, the younger generation needed a release from being locked down and on occasions, this led to destruction of urban bushland areas. This occurred all over the suburbs.'

But Crag and his team are trying to engage the community. When we talked the council had recently employed a nature engagement officer: 'She's been brought on specifically to speed up and expedite some of the projects in the Yarra nature strategy, particularly the gardens for wildlife program, which is working with residents looking to encourage wildlife back into their gardens ... plus a range of other projects mainly around nature engagement and promotion, trying to get the community understanding and appreciating ... the wildlife that exists here. After all, we protect what we love!'

From owls to gliders to fruit bats to snakes to everything else, we're extraordinarily lucky in the City of Yarra to be in the inner city and live alongside so much wildlife, so much nature. It's one of the reasons why I've lived here for the entire time I've been in Melbourne. Even after nearly twenty years I still see animals here that surprise and delight me, that I've never seen locally before. And the thought of what wildlife might be still to come is truly exciting.

POWERFUL OWL

It was a lockdown – who can remember which one – and I was at home on the computer becoming increasingly obsessed with the sugar gliders of the Yarra corridor. I found a seminar online, hosted by Melbourne Water and recorded a few months earlier, given by a man whose job it was to put up nest boxes for the gliders. At the end of the talk, in a throwaway comment, he mentioned that there were sugar gliders in my local park. News to me! There was a pretty decent patch of bush in part of the park only a kilometre from my house so in one of my hours of allowed daily exercise I jumped on my bike, pedalled my guts out (counting every minute, eager not to waste a single one) and went for a wander, looking for nest boxes. I found one so I returned that evening or the next to stake it out. I watched and I watched and I watched from sunset until it was dark and not a single thing emerged. I'd had a pretty good idea that would probably be the case: there were spiderwebs in the entrance hole. Still, you have to try.

What I did notice, though, was a strange double hoot coming from the far side of the river. It repeated, over and over again, the first note slightly longer, the second note slightly clipped and usually higher than the first. It sounded a little bit like a dove and a little bit like a boobook owl. I recorded the call on my phone. After a while it stopped, but a large bird flew silently into the trees and landed just beyond the reach of my torch and glared at me: I could see its eye-shine but not much else.

Later that night I uploaded the recording to a Facebook group full of people who are expert at identifying birds and straight away someone had the answer: 'Female powerful owl'. Well I forgot about sugar gliders pretty quickly. Been there, seen that (sorry, gliders, I still love you). A powerful owl! So close to home! *That* was exciting. And god knows lockdown needed a bit of excitement (not anxiety, no shortage of that but it's not the same thing).

Now I've seen powerful owls before. There's a well-known pair just a forty-minute bike ride up the Main Yarra Trail from my home, in another big bush park on the banks of the river. There's another well-known pair much further out, beyond the trainline and beyond the bike path, so harder for me to get to. And there are frequent sightings between these two pairs, all up and down the river, and even downstream: a few years ago there were powerful owls hanging out in the Royal Botanic Gardens in the city. They've also been spotted in Carlton Gardens and Fitzroy Gardens, within sniffing distance of the CBD. One was even photographed outside a well-known bookshop in the middle of Carlton. They're out there, lurking in trees and pouncing on possums. They like to go to sleep with the remains of last night's prey under their talons, waking up to have a little nibble from time to time then going back to sleep, like someone conking out after a big Saturday night with their midnight kebab clutched to their chest. They're surprisingly easy to spot, too: if the spatter of white bird poo and pellets full of bones and fur on the ground doesn't tip you off to the fact that Australia's largest owl is in the tree above your head, the sight of a massive bird shaped like a nightclub bouncer glaring down at you through the branches soon will. They're not exactly inconspicuous; you've just got to be in the right place. The problem is that there is an unfathomably huge number of places, and nearly all of them are the wrong place.

So I felt pretty bloody smug that I'd just stumbled upon a powerful owl hanging around in my local patch of bush. I really can't claim any credit: she wasn't exactly being shy. I guess when you're an apex predator who eats possums for a living you can afford to not be too shy. I took to returning to the vicinity of that empty glider nest box just about every night, and just about every night the owl would hoot: 'Hoo-*hoo!*', repeated every seven or eight seconds, starting a few minutes before sunset and stopping just before the possums woke up half an hour later. She'd call from the far side of the river, take a pause, then fly across and start up again on the near side, and I'd scramble along the rutted dirt tracks through the bush, hoping to see her before she went silent and disappeared. Very often I did. I'd seen powerful owls roosting before, but seeing this huge, predatory nocturnal bird in her element, after dark, being active, calling out, flying around, was a completely different experience. It felt like I was seeing a powerful owl for the first time, *really* seeing it.

She had certain perches she preferred. One was the bare branch of an old red gum sticking out over the river, and she'd sit there and twitch her head at the fruit bats circling down to drink from the river. Whenever she hooted she'd puff out her chest, and open her wings a little, like an opera singer lifting her arms to sing to the back of the room. She'd see me long before I saw her, every time, and she knew when I was watching her: more than once I'd turn away for a split second, and when I turned back she'd be gone.

I never heard her go, of course: when they fly owls are completely silent. Sometimes I'd see her fly off, and she'd disappear up into the bush without making a sound. Or I'd see her at a distance, flying around, and I'd assume that she'd just gone from one branch to another in the same tree and then I'd hear her calling from a couple of hundred metres further back into the bush.

The thing about powerful owls is that they aren't just silent; they're invisible, when they want to be. I remember years and years ago when I was a kid one of my friends who was into military stuff and combat and stealth and all that kind of thing (we all had a friend like that in high school, didn't we?) telling me that the best colour for creeping around at night isn't black, but dark green or dark brown: most of the time the night isn't black, not really, so something that's black can be seen, if you're really looking, because it's *darker* than dark. Powerful owls are dark brown, and when I was watching my local powerful owl I noticed how she'd fly around just under the canopy, in the shadows, and her plumage blended into the dark *but not black* shadows of the bush at night, and she'd just ... vanish. The perfect stealth hunter.

Apart from the hooting. I was surprised, often, to see possums clambering around in the treetops only a few metres from where the owl was hooting, but reflecting on this I suppose that if the owl's being vocal then they know exactly where she is, so maybe in a funny way that's the safest time to be a prey animal hanging around near your major predator. Keep your friends close, and all that. Anyway she hooted a lot, and with good reason: the days were getting colder and shorter, the start of the year was turning into the middle of the year and powerful owl breeding season was beginning. They breed through the coldest, darkest months, and raise their chicks when the year comes out the other side; if everything goes well white, downy chicks can be seen outside the nest hollow in September and October. My local powerful owl was hooting not only to announce her territory, but also to attract a mate.

And one amazing night, just as I was turning to head back home, a male powerful owl hooted back at her.

His voice was slightly deeper than hers: unlike most owls, male powerful owls are larger than female powerful owls. I watched him as he hooted, and glared at me, just as the female had done so often in the weeks before he appeared, and in the distance I could hear the female, faintly, hooting as well. No doubt about it: two owls, one high, one low. A pair, in breeding season, in my local patch of bush only a kilometre from my house on the banks of the beautiful, nurturing Yarra River.

Pretty soon they found each other, and I saw and heard them both a bunch of times. When lockdown ended I took a friend, well-practised at keeping secrets about sensitive wildlife, to see them. We sat at a respectful distance from where the female owl crossed the river every night, and we saw her fly in and land, and then we saw the male land as well, and they sat side by side, calling softly to each other. The male powerful owl prepares the nesting hollow; it's up to the female to decide whether it's a worthy home. We couldn't decide if they were a mated pair—

And then. And then! We saw the male *disappear inside the tree!*

It's very hard to completely lose your shit in excitement and also stay completely silent at the same time. It's even harder when you're sharing the excitement with a friend. In the night-time gloom under browny, overcast skies my friend and I stared at each other, eyes and grins equally wide. As we turned our attention back to the owls the male re-emerged from his tree, and then ... and then I can't remember what happened. It's possible I blanked out.

Needless to say, I kept going back. I stayed what I thought was a respectful and respectable distance from the tree but even so one night I saw the male glaring at me from a low branch, and then, as had happened before, I turned my eyes away for a split second and when I turned back he was gone. But then I suddenly saw him again, a massive silhouette in the gloom – and he was coming straight towards my head! I'm not too proud to admit that I hit the deck in a spectacularly undignified fashion as the male flew right over the top of me and landed on a branch 10 or 20 metres away, where he proceeded to glare at me some more and bob his head in agitation. Okay, mate, point taken. Too close to the nest.

I took to watching the tree from the other side of the river instead. On a couple of occasions I saw both male and female fly out of the hollow like they were shot out of a cannon. Once I even saw the male take a pass at a brushtail possum in the outer branches of a tall river red gum right above my head: the possum made a narrow escape when the owl got his wings caught up in the twigs. If having one powerful owl to watch was thrilling, having two – and a pair! – was exhilarating. The possibility of chicks in a few months was nearly all I thought about.

But something wasn't quite right. It turned out the male was roosting nearly a kilometre away, every night: he should have been staying close to the hollow. And after the first flurry of excitement I never saw him visit the hollow; in fact I didn't see any activity there. One night I brought a fold-up chair and a thermos of soup and several layers of warm clothes and I sat there, on the opposite bank of the river, for three hours, watching, and nothing happened at all. When I finally gave up and was riding home something very like a powerful owl, huge and dark and silent, flew over my head, heading away from the hollow.

There had been rolling lockdowns all through the start of the nesting period. There was a walking track just 20 metres from the hollow and every time there was a lockdown the track was heaving with people: mountain bikers with lights as bright as comets, walkers chatting loudly to each other or walking their dogs, joggers huffing and puffing. On the river in front of the hollow canoeists paddled and splashed noisily back and forth. Everyone eager to make the most of the limited amount of time they were allowed to be outside – and after all, I was the same, the only real difference between them and me was that I knew there was a pair of powerful owls nesting *right there.*

Or at least I thought there was. Probably. It was getting less and less certain all the time. And then one day I went to check on the male at his isolated roosting spot, and he wasn't there. And he persisted in not being there. He wasn't near the hollow, either. And then, finally, I was watching the hollow again and for the first time in a long time I heard the female hoot – from the far side of the river, on the opposite bank to the hollow, from where I'd first heard her call.

I suspect they were new to it. I think inexperience and disturbance was an intolerable combination for them. I don't know what happened to the male: he was seen by someone in Clifton Hill, next to Merri Creek, and that was the last I heard of him. I went back at the start of the next breeding season and the female was still there: still calling, from the same place as before, and still perching in the tree where the hollow is. One time I was looking for her and she flew across the track only 2 or 3 metres in front of me and then down the slope beyond so low that I could see how she held her wings as she glided, slightly bowed from the shoulder. I've been watching her for two years now – as I write this – and she hasn't found a partner in the second year. Maybe by the time this book comes out she will have.

But she's still there, still the empress of her domain. Still terrorising the local possums – and there are a lot, in that patch of bush: once on a forty-minute walk I counted thirty-five brushtails and twenty ringtails. She's still only findable when she wants to be found; still capable of suddenly going silent, flying off under the canopy and disappearing. Still such stuff of legend that it's almost impossible not to feel a little mystical about her. In one week early in the 2022 breeding season I found, in the space of three days, the remains of two ringtail possums left on the street right outside my front door, remains which were characteristic of powerful owl kills. I'm not being fanciful when I say that she may very well have been perched on my rooftop while I was asleep, and I'm still wrapping my head around that. And I'm probably imagining it but I feel like she's got used to my presence: we've locked eyes on so many occasions now, and she hasn't flinched, hasn't head-bobbed, hasn't fled – just grown bored of me, as only an empress can, and turned away to focus her attention on some other thing.

Oh, and one last thing to note. One cold night in 2022, keeping an ear out and spotlighting in that patch in the forlorn hope of stumbling across her, I saw two tiny pairs of eyes peering out of a narrow vertical hollow in a dead tree: turns out there are gliders there, after all.

WE DON'T THINK LIKE THAT

'What we've been talking about,' Frank told me, continuing a conversation that had begun forty minutes earlier in his car and was now being conducted on foot, 'is seeing the river as a single entity in its own right, and once you start looking at it like that you realise what a fantastic asset it is, you see the river in a completely different aspect.' We were walking around a small reserve in Bend of Islands, a short drive from Warrandyte in Melbourne's outer suburbs, along with Frank's friends Ross and Liz. 'Anyone can do that whether they live on the river or they live in Dandenong or somewhere, it doesn't matter. Once you see it like that and that it's worth preserving it changes ...'

If I hadn't been told that the area we were walking in was a reserve, I wouldn't have known. It was my first time in Bend of Islands, and I couldn't discern this part of the bush from the rest of the bush that surrounded us, interrupted only by dirt roads and the occasional house. My eye was untrained. But I was the guest, and I had guides who'd been living here for decades and who knew the land's every nuance.

I explained to them all that I wanted to celebrate the Yarra, to reverse the negative attitudes that so many people still have towards it. 'We don't think like that,' Liz said, before adding with mock wonderment: 'But yes, I recall that people did this!'

When we came to a patch of recently turned soil, bare and brown amid the leaf litter and groundcover all around it, Ross was unhesitating: 'Wombat. We've got a wombat that's running around our house at the moment, we haven't had one for ages.'

'We get scratchings of echidnas all the time,' Frank added. 'Lyrebirds.'

'You'd see holes, wouldn't you?' Liz pointed out. 'If it was an echidna.'

'Yeah,' Frank confirmed, 'when the scratching's from an echidna there's a pointy hole.'

They debated the location of what they called the Wedding Tree – 'His daughter got married,' Liz explained – and whether we had time to walk to it. We were due back at Ross's house for afternoon tea. 'The candlebark?' Frank said. 'That's *way* down.'

Soon we came to a patch of common bird orchids sprawled across the ground. 'Oh, there's heaps!' Liz exclaimed.

'Frank spends a lot of time here monitoring orchids and identifying [them],' Ross told me.

'Bird orchids typically have hundreds of leaves,' Frank said. 'This species has a lot of flowers but there's a couple of other species here where you get hundreds of leaves and no flowers.' After a few minutes of searching we found one opened flower. 'Yeah that's open, that's it,' Frank said, excited. 'So it looks like a bird. And see the yellow? That's the pollen. So it attracts insects that go in there and when they come out they put the pollen on themselves and they go to the next one and pollinate. Down a lot further ... there's another orchid called the autumn bird orchid which is an absolutely beautiful orchid. It very rarely flowers, two flowers there's been this year.'

Walking down the hill we came within sight of the river – 'I think we can see it enough,' Frank said, mindful of the time – and I asked them about weeds. 'Years ago we were inundated with basket willow,' Ross said, 'and they were clogging up the river. Our son-in-law who's a botanist, he poisoned them and they died and they fell apart like that. It was just amazing, they disappeared and they never came back. And the other side [of the river] which was the most inundated area, it's just fantastic how the growth [of native vegetation] has just suddenly come back.'

Frank indicated an expanse of grass in front of us. 'All this stuff here. This is all *Ehrharta*,' he said. 'It's really beyond control ... we get grants from the council to do specific things, and one of those is the treatment of the bridal creeper and the boneseed around the river.'

'There's also some wandering trad problem,' Liz added.

Frank said: 'Melbourne Water actually now have a contractor who does an annual run along the river as well ... it's a fantastic improvement on what it was forty years ago. But the island itself, which we haven't really talked about, it's only quite a small island but forty years ago that was covered in blackberries and Melbourne Water cleaned it all up.'

Not every change has been so welcome, though: 'Twenty years ago there had never been a deer sighted in Bend of Islands,' Frank said, 'and now they're just ... like the gullies I weed up near the Co-op, I used to have to fight through this dense vegetation. Now there's a superhighway through every gully. It's a dreadful problem ... they're incredibly intelligent animals.'

We crunched across the gravel to Cric and Ross's mudbrick

house, surrounded by an indigenous garden that blended into the surrounding bushland. Inside the house, we sat down to cake with Cric and Janet. I asked them all how they imagined the future of their secluded corner of the Yarra.

'We like to think,' Ross began, 'with the energy of the people that are here, both now and in the future, the more it goes on I think more and more people are convinced that it's special so you attract people that are enthused about it, they want to do things to make sure it's maintained.'

Janet spoke up, quietly and thoughtfully: 'I think that still for the next ten years a lot depends on us, who have been here [for decades] and who understand it and are able to pass on our knowledge to the new people coming in and enthusing them to still be part of the culture of why we're here. At the moment our group is dwindling and we're not pulling in the new people fast enough, I think we're at a bit of a turning or crisis point.'

I expected the turnover rate of houses in Bend of Islands to be slow, but the group told me it was equivalent to the rest of Melbourne. The area was previously owned by Melbourne Water, Frank told me, and was scheduled to be dammed in the 1970s, until an environmental assessment found that it was an area of special significance. Instead it was designated as an Environmental Living Zone, and residents started to move in, acquiring land from Melbourne Water. Frank and Janet live on 'the Co-op', the Round the Bend Conservation Co-operative. 'The phrase [round the bend] is ironic,' the Co-op's website says: 'In the early 70s when the Co-op was formed, the idea that people would live in the bush to protect it, rather than exploit it, was seen as a bit crazy.'

'I think the strength of this area is really helped by having the Co-op,' Ross said, thinking out loud, 'because the Co-op's got a much tighter approach, there are fewer people, they all look at each other much closer like this. As a cohesive group within the big group they are a strong heart, the Co-op, for the whole Bend of Islands. If we were just individual block owners right through the area we wouldn't have the same spirit, I don't think.'

Getting back to the topic of the future of Bend of Islands, Frank said: 'The problem with community organisations, as you probably know, is that as long as there's someone doing something no one else will put their hand up for it, you have to walk away and make a vacuum and say, "They're gone, shit, what are we going to do now?" and then someone will say, "I'll take it on and try and learn."'

'If there's a crisis in this area, an existential crisis for this area, and there's no one to fight, and they get motivated, they'll fight and they'll win and then they'll be where we are now and we'll be right again for another twenty years.'

We got to talking about the wildlife in Bend of Islands: earlier Frank had shown me a photo he'd taken of a white-throated nightjar, a superbly camouflaged ground-roosting nocturnal bird related to tawny frogmouths. He'd recorded it breeding – the closest to Melbourne this species had been recorded breeding, he told me. He also showed me photos of brush-tailed phascogales, and told me about Bend of Islands' lyrebirds: the Bend of Islands Conservation Association [BICA] website explains that:

> Lyrebirds had not been recorded in the Bend of Islands or Warrandyte State Park for at least 38 years, (since before the '62 fires), until they were seen in the Bend of Islands in July 2000. After 270+ records of the bird in the Bend, in May 2015, they were finally recorded to have crossed the Yarra into the Mt Lofty section of WSP [Warrandyte State Park]. There are now over 500 records.
>
> In 2020 breeding birds were monitored and evidence of previous breeding, from 2015 or before, was found.
>
> This re-establishment of the Lyrebird in WSP is a practical

> example of the importance of the Habitat Link [provided by Bend of Islands] between Kinglake National Park [KLNP] and Warrandyte State Park. WSP is too small and fragmented to independently sustain its biological diversity in the long term. The Habitat Link with KLNP is a way of overcoming this problem. The lyrebird experience is concrete evidence of the importance of the link.

Other parts of Melbourne benefit from this connection too: Bend of Islands acts a corridor by which wildlife can move into Melbourne. Frank, Ross, Janet, and Liz described it to me as 'like living in a national park'; with no domestic dogs or cats allowed in Bend of Islands, and no non-indigenous planting, wildlife is able to move in or move through the ample habitat provided with a minimum of disturbance or risk. But Bend of Islands is also a biodiversity hotspot in its own right: BICA's website lists nine species of amphibian; fourteen species of reptile; fifteen species of mammal ('an incomplete list') and 181 species of bird as having been recorded within Bend of Islands – not to mention 40 of Victoria's 130 species of butterfly, and numerous other invertebrates. BICA's list of plants indigenous to the area runs to eleven pages. It's a richness of life of which the residents of Bend of Islands are rightly proud, and protective: Frank, Janet, Ross, Cric, and Liz were eager to impress upon me that the Bend of Islands community has a custodial responsibility to provide ongoing research and education about the special and precious area that they call home. They see Bend of Islands as nourishing the Yarra River ... giving it a 'biodiversity recharge' as it starts its journey through Melbourne.

'We have a powerful owl here every night,' Ross said matter-of-factly over our afternoon tea.

'They've been breeding in this area for years,' Frank confirmed.

I asked them if there were bandicoots in Bend of Islands. 'No, we haven't got any records of them,' Frank said. 'That's not to say they're not here, in fact they found a bandicoot up in Christmas Hills [next to Bend of Islands].' He related a conversation with a Melbourne Water employee who'd been surveying the area around Bend of Islands: 'He said [the area where they found the bandicoot in Christmas Hills] is just trashed. He said, "You guys should be doing surveys down here because it's highly probable they're there." So ... just in the last few months we put up some camera traps to see.'

Frank then related a story: 'The finding of these rare things is really quite serendipitous,' he said. 'I was walking [around here once] and I saw this jewel beetle, and I said, "Oh that's a jewel beetle, I'll take a photo of that," sent it off to my mate who's into jewel beetles, he sent it off to the expert, and he rang me up and said, "The expert's trying not to cry. He's been looking for this species for forty years and he's never seen it."'

On a roll, Frank continued, relating an encounter with an American entomologist: 'I said to him, "Is it the same in America as it is in Australia that only half the insects are named?" and he said, "No, no, everything's named in America, pretty well," and he said they do find species, but pretty well everything's named. And he said in Europe, if you find a new insect species, they'll carry you down the street on their shoulders, that's how important it is. And here we are, you can walk out into your back yard, pick up a rock, see an insect and there's a 50 per cent chance that it's not named. There's a push by the conservationists at the moment to say we need to inject serious money to get all these things named. We need to be ... training special teams of young biologists and botanists to get out there and investigate all these unnamed species and name them and see what their properties are, because we're losing them faster than we can name them ... No political will, that's the issue. On one hand everyone says Australia's environment is unique and fantastic, but on the other hand they say, "But we're not spending any money on it. We're just going to rip it apart." So it's a difficult situation.'

'Anyway,' Janet said, 'we try. In Bend of Islands, we try.'

BRUSH-TAILED PHASCOGALE

I sometimes find myself torn, when writing, between assuming that anybody who I'm lucky enough to have read my work is knowledgeable about wide and varied subjects, and realising that I cannot reasonably expect everyone else to have the same combination of fascinations and collected facts as me; which is to say I sometimes find myself torn between a fervent desire to not condescend to any reader and an acute awareness of the egotistical folly of believing my own interests to be universal; which is also to say, in a more direct and immediately relevant way: forgive me when you read this next part of the book if you already know what a brush-tailed phascogale is, but I'm aware that many readers may be thinking at this point, *What the bloody hell is a brush-tailed phascogale?*

Let me begin answering this question by suggesting that perhaps you know what a dunnart is, or an antechinus, or – in increasing order of likelihood – a quoll, or a Tasmanian devil. The latter two, famously, are carnivorous marsupials; the former two, a little less famously, are also carnivorous marsupials, but of smaller size. The brush-tailed phascogale, as you may by now have guessed if you didn't already know it, is also a carnivorous marsupial, of a size intermediate between the two aforementioned pairings.

Or to put it more succinctly: it's a marsupial that hunts and eats other animals and it's often described as being about the size of a rat.

I have no wish to disparage rats, of which you'll later learn I am quite fond, but let me clarify that aside from size, and a few other traits such as a pointy snout and general mammalness (fur, four legs, milk ducts), the brush-tailed phascogale is entirely its own thing (or not entirely, because there's also a red-tailed phascogale, but they live in Western Australia so we're not dealing with them here). It lives in box-ironbark forests, of which there are many fewer than there used to be, and it scurries up and down tree trunks at night, lifting loose bark with its pointy snout to gobble up any small animals found underneath, and occasionally leaping from the base of a tree to the ground, where it bounces after terrestrial animals or up neighbouring trees. It reacts quickly and with fear to the sound of powerful owls, flipping to the far side of any tree it's on and listening intently, but when satisfied that it's safe it reverts just as quickly to hunting, moving in jerky, staccato motions up and down the tree trunk, grabbing onto the bark with its sharp claws.

Or at least, all of this is what the brush-tailed phascogale that I saw on the forested banks of the Yarra in outer Melbourne in early 2021 did.

I'd seen photos of phascogales in the exact same area, and had resolved to investigate for myself. A friend with whom I often go exploring in the bush (actually the same friend I watched the powerful owls with) was of the same mind. I had a day off work so, having consulted my copy of Barbara Triggs' *Scats, Tracks and Other Traces*, I caught the train to the nearest station then rode my bike to the river, where I investigated every fallen log I could find, looking for phascogale scats that would indicate a latrine and thus the nearby presence of the animal. Scats found, my friend and I returned at night and staked out the latrine log, each of us armed with a red torch – red light being less disruptive to nocturnal mammals but particularly phascogales, or so I'd read, because they have a habit of hiding from white light.

We waited until it got dark. We waited until well after it got dark. Possums usually emerge about half an hour after sunset in

Melbourne – but at possum o'clock there was no phascogale. I turned my red torch on from time to time, looking along the beam for eye-shine, but there was nothing. Winged animals teased me as I sat: I heard the distant trill of juvenile powerful owls, and I'm pretty sure one flew past me in the gloom, vast and dark; I know for a fact that mosquitoes fed on any patch of skin I left exposed. I would've preferred it if they hadn't.

And then! I switched on my red torch for what felt like the hundredth time, shining it along the length of the latrine log as I had done before, and before, and before, but this time there was something unmistakeable, something always thrilling: the bright red eye-shine of an animal looking back at me. Red torches make everything spooky, and I'm sure they confuse and unnerve anybody who just happens to be passing by and sees blood-red light illuminating the bush. One time my friend and I went spotlighting in the Mallee, in the darkest dark, and our red light fell upon the skeleton of a kangaroo encircled by viciously sharp spinifex and the effect was positively demonic. But in my experience, apart from not upsetting nocturnal animals, red torches make eye-shine stand out even more strongly than white torches: looking along the torch beam, seeing what it sees, the red light illuminates everything dimly, and in that light an animal's eyes shining back at mine are like twin moons dazzling in the night sky.

That's what I saw at the end of the latrine log. And behind the eyes, I saw, was a long body and four legs splayed out, gripping the trunk of a small tree; and at the end of the body was a black tail like a bottlebrush, twitching in irritation as the animal saw or smelled or heard me. And then it was gone, scrambling vertically down the tree and diving into the grass.

I sent my friend a frantic, urgent message. I could see on my phone that the message went unread. She'd wandered off to spotlight elsewhere; I could see her torch beam and I whisper-yelled her name hoarsely to try to get her attention. I'd forgotten all about the mosquitoes. I didn't know if I'd be able to find the phascogale again: this small, fast animal moving in an unknown direction through shrubby understorey and hundreds upon hundreds of trees. I stood up and scrambled my way through the bush towards my friend, calling her name as loudly as I dared.

I finally caught up to her and we set off back in the direction of the log. There was a dead-end dirt road about 10 metres from the log and I thought the phascogale might have gone in that direction, so we set off. I'm probably remembering this incorrectly; I was too excited to commit those moments to memory perfectly, and thinking back on it now my memory is dominated by what came next. My friend and I went to the road, and we walked down it carefully, slowly, and then we heard the noise of bark being torn up. We shone our torches (it probably wasn't anywhere near this exciting; or it was definitely ten times more exciting than this) and there it was! Or at least let's say it was there because, however it happened, the phascogale turned up sooner rather than later.

I know for a fact that my friend and I wheezed in disbelieving, whispered laughter as we watched the phascogale's antics. I know this because I recorded those antics on my phone, and I still have the video, and the video has our voices on it. We stood on the road and watched the phascogale as it raced up and down tree trunks, voraciously forcing its wedge-like snout under loose bark and ripping it up. It leaped from a tree into the long grass, and we saw it grab and devour some huge insect there – a praying mantis, perhaps? Or a dragonfly, such as one we'd seen sleeping on a bush the previous time we'd visited that spot? It flounced – there really is no other word for it; if you've ever been to the northern hemisphere and seen a squirrel on the ground you'll know the movement I'm describing – as it moved through the long grass, not running but leaping in high parabolas. I have that on video, too. I watch it a lot, and I listen to the delighted laughter of my friend and me and I smile all over again.

We were watching this whole performance under red light, but at one point my friend asked if I could put a white torch on the animal, just for a second, just long enough for her to get a photo. We knew it would only be a second because the one thing every guide and website and knowledgeable wildlife watcher says about brush-tailed phascogales is that they hide from white torchlight. So my friend lined up her camera, got the focus right and did whatever else it is photographers do, readying herself for a split-second chance; I asked her if she was ready and she confirmed that she was; I switched my torch from red to white ...

... And the phascogale ignored it! Behaviour: unchanged. Rat-sized carnivorous marsupial: unbothered. Rulebook: chucked out. It kept doing exactly what it had been doing, scampering up and down trees, gobbling up invertebrates – and all only 5 metres away from us. We were on the road and the trees were 50 metres deep at least and the phascogale decided that right there, on the outermost row of trees, under a white spotlight, right next to the road, definitely within hearing and seeing and smelling distance of two near-hysterical humans, was where it wanted to be that night. We weren't about to argue with it, or try to persuade it otherwise. It waved its tail at us once, as it had done when it first saw me, a gesture that is speculated to be a sign of annoyance, though nobody is really sure – but if there's one thing I've learned about watching nocturnal animals, or just about any other kind of animal, it's that torch or no torch they're very aware that you're there and if you're lucky they'll just decide to ignore you. By that criterion I don't think I've ever been luckier than I was on the night my friend and I saw the phascogale.

We watched it for half an hour. We could hear traffic and trains only a few kilometres away at most. We were in the suburbs of Melbourne, having what we both agreed then and since was one of the single best wildlife encounters of our lives. Both my friend and I have had a *lot* of wildlife encounters.

There aren't many nocturnal mammals in that patch of bush. It's young bush: there are very few old trees, and so few hollows. But phascogales don't need very large hollows; they don't take up much space. And there aren't too many around – they're listed as vulnerable in Victoria. But a few of them are still at home on the banks of the Yarra, living fast, short lives, eating anything they can catch.

MANY PEOPLE PUTTING TOGETHER

It's a little embarrassing to admit that even after nearly twenty years of living in Melbourne, and living alongside the Yarra for nearly half that time, before writing this book I'd never been up the Yarra Valley. Sure, not having a driver's licence didn't help – though it turns out you can get buses up the valley. Tourism in the Yarra Valley, or at least its marketing, is heavily directed towards wineries and winery tours, and as much as I love the odd glass that's not really my thing. Every time you see a photo of the Yarra Valley splashed on the side of a train station or popping up in the middle of a website it's of misty vineyards, green grass, rolling cleared hills. The Yarra Valley seems ... agricultural. And look, I love eating and drinking so no slight against agriculture, but I don't go on holiday to see farms. Again, just not my thing. Some people love that, and that's great, but it's not for me.

So it wasn't till I was writing this book that I thought, *Well the Yarra's not just an urban river, I'd better go check it out.* This might sound pretty naive to those of you who are a bit more open-minded than me, so if you find your eyes glazing over, feel free to skip this chapter. Though there's some pretty interesting stuff in here, I reckon.

I'd thought about going camping with my girlfriend but life got busy so that turned instead into a daytrip, which then turned on the day into more of an afternoon trip, but that's okay because it turns out the Yarra Valley isn't very big. Or more accurately, it isn't very *long*. From the outskirts of Melbourne to the official start of the Yarra River is only about 65 kilometres in a straight line. From the start of the river to its mouth is only 110 kilometres (which tells you as much about the urban sprawl of Melbourne as it does the length of the Yarra Valley). Of course, rivers don't flow in a straight line, and certainly not the Yarra: the river's total length is actually 242 kilometres.

It was a Saturday morning in January 2023 when we set out, my girlfriend and I, up the valley. First stop was lunch in Yarra Junction, then on through Warburton: our vague plan was to drive as far as the Upper Yarra Reservoir, check it out, then work our way back down, stopping along the way.

The Upper Yarra Reservoir was opened on 26 November 1957, a century after Melbourne's original reservoir – Yan Yean, which diverts water from the Plenty River, one of the Yarra's major tributaries. There's a big plaque on top of the dam wall at the Upper Yarra Reservoir that tells you all this. It was a warm and sunny day when we stood on top of that wall, and the water on the other side of the fence was clear and green; we wished that we could dive in to cool off. But you can't do that, of course: the reservoir provides most of Melbourne's water. Forested hills rolled out layer after layer as far as we could see, and clouds of tree martins buzzed around our heads in the sunshine. The reservoir has a total capacity of 200,579 megalitres. To view it you walk across the top of a monumental spillway, a vast concrete chute at the bottom of which the river – what of it isn't held in the reservoir – flows.

After the heat of the sun got the better of us we drove back down the zigzagging road to the foot of the dam wall, and the riverside, where large, open grassy areas provide camping space. It was the tail end of the school holidays and a scattering of people were camping there. An overgrown cricket oval stretched out at the bottom of a short flight of stone steps; a hall formerly used by dam-construction workers stood locked up. A restored waterwheel from before the dam, a remnant of when there had been a colonial settlement here, stood amid long grass by the side of the road.

My girlfriend and I had planned to do some bushwalking, but as soon as we found a track we found, too, that it was blocked off: a wooden bridge built on top of a fallen tree was closed with yellow tape, presumably in the aftermath of recent floods and heavy rain. So instead we contented ourselves with stepping carefully down the riverbank to the river which in this location, only a few hundred metres below the dam, was not much wider than a creek. It was rocky, and swarms of tiny black insects rested on every rock that wasn't submerged in water. In the still sections, the backwaters, the rocks below the water were covered with algae, but in the middle of the river where the water flowed the rocks were grey and clean, and the water was as clear and as clean as any river I've ever seen before. I cupped my hand beneath its surface, felt the soft coolness of it flow effortlessly over my skin; watched my fingers wobble beneath the surface like an old analogue TV losing reception. The water was so clear that the line of that wobble, above which my hand's outline firmed up, was the only indication of where the boundary was between water and air. I lifted my hand out of the water and splashed it over my sweaty head; did it again, and again; rubbed the water through my hair. My girlfriend took a photo of me wearing a grin that was so big it nearly split my face.

'In the upper reaches of the Yarra you probably know it's a pretty clear waterway,' Sarah Gaskill told me a couple of days later, before I'd even mentioned to her that I'd been up at the top of the valley. Sarah works at Melbourne Water – Charlotte Sterrett, the Yarra Riverkeeper, had put us in touch. 'As the river travels down the valley, it changes it's nature,' Sarah said.

On the drive back down the valley my girlfriend and I stopped just outside Warburton, only about 20 kilometres from where I'd placed my hand in the river. The river there was still clear, and sandy on the bottom. Forest rose on either side, tall eucalypts and tree ferns, many other plants growing thick and lush. The sun filtered bright green through their leaves but the forest and the river cooled the air. Under the blue sky the river sparkled, and downstream from where we stopped we could see that a fallen tree had created an impromptu dam: the water before the tree was wrinkled with speed, but the water on the other side was smooth and flat, lulled to gentleness. Not that the river here felt anything other than gentle, really; it was clear and shallow, and I imagined I'd be able to see fish in it. I looked closer – and saw, to my dismay, that even here, so close to the catchment where the public aren't allowed, there were a few pieces of rubbish submerged beneath the water, draped around twigs and snags.

If I had looked for long enough I might have seen fish,

too, though. Sarah told me when I spoke to her two days later that 'when you get to the upper reaches there's some cracking habitat for blackfish'. She mentioned Australian grayling, too, and pointed me towards the Native Fish Report Card (NFRC). The NFRC program began in 2017 and targets three species in particular in surveys of the river: Australian grayling, Macquarie perch and Murray cod. The latter two have both been translocated into the river, but the Australian grayling has always been in the Yarra. It's an endangered species in Victoria and a threatened species nationally, and the latest report, from 2022, notes:

> Recruits were detected for the first time in 2022, while juveniles (2018, 2019, 2021 and 2022) and adults (2018 and 2021) have been recorded in low abundances ... The presence of recruits in 2022 and juveniles in 2018–19 and 2021–22 indicates stream conditions were suitable for recruits to be attracted into the system.

Good news, though offset somewhat by the fact that, as the report makes clear, only five individuals were caught in the 2022 survey. As the report notes: 'The NFRC does not expect to capture enough Australian grayling to measure key health indicators. Rather, threatened species such as Australian grayling are targeted for monitoring to gain a greater understanding of the current status of the populations which is essential information to inform the management of such species.'

Besides the grayling, the perch and the cod, other native species that have been detected in NFRC surveys since 2017 are Australian bass, golden perch, river blackfish, short-finned eel, tupong, barramundi (like the perch and the cod, a translocated species), Australian smelt, common galaxias and flathead gudgeon. Native fish that are known to occur in the Yarra but have not yet been detected in the surveys include climbing galaxias, dwarf galaxias, ornate galaxias, spotted galaxias, pouched lamprey, shortheaded lamprey, southern pygmy perch and Yarra pygmy perch. Surveys have, though, recorded southern Victorian spiny crayfish, another threatened species.

'The negative story,' Sarah told me, 'is about the carp and their role in the waterway – you're probably aware they're quite an invasive species called "the rabbits of the waterways", they just turn everything muddy and brown and just fossick around on the bottom of a river. With the floods we saw them coming into the billabongs so they just munch up the bottom of the vegetation, which is why a lot of our billabongs aren't as clear as they could be ... Carp are the big challenge.'

Some native fish species in the Yarra, such as the Australian grayling, are migratory, moving between fresh water and salt water, and Dights Falls just below the confluence of the river and Merri Creek had been an obstacle to their movement upstream from the lower Yarra. A fish ladder was built in 2012 to help with this: a series of chambers fill with water at gradually increasing levels, allowing fish to move from the bottom of the falls to the top. I asked Sarah whether the fish ladder at Dights Falls was having a beneficial effect. 'Oh without a shadow of a doubt, yes,' she said emphatically. 'There are barriers along the river ... and the Dights Falls [fish ladder] is really helping.'

Helping out the grayling is part of Sarah's job description: she works in the part of Melbourne Water responsible for providing environmental releases of water into the Yarra on behalf of the Victorian Environmental Holder. 'Broadly speaking, the Yarra has a bucket of water put aside every year – so it gets seventeen gigalitres of water on the first of July – and our role is to make sure that that water gets delivered to the river for the river's needs,' she explained. 'So how we go about doing that, is we do what's called an environmental flow study. So we split the river up into similar reaches, so you can imagine the upper Yarra is different to the mid-Yarra to the lower Yarra, so each of those reaches has a different watering requirement, different flows that perform different ecological functions – whether it's providing a flow for the grayling to come up and spawn, or whether we're providing a boost to move silt through for maybe the blackfish to spawn, or whether it's supporting vegetation in the banks in stream, et cetera.'

When you get to the core of it, though, this is giving the river its own water back. Of the seventeen gigalitres of water put aside every year, Sarah says: 'We don't think that's enough for the river to be healthy.' There's ongoing monitoring of the river's health, and how the environmental releases affect that. The environmental release program has only been going for eleven years, and being a river it's a dynamic system: some years will be wetter than others. When I spoke to Sarah it was only a couple of months after rainfalls so enormous that Dights Falls had disappeared completely under floodwater. I asked her how climate change, with its predicted increase in extreme and unseasonal weather events, might impact the river and its ecosystems. She spoke again of the Australian grayling, which spawns in fresh water and requires seasonal flow to migrate down the river into the Bay: 'So having those large flow events out of those breeding seasons would certainly, or potentially, have an impact on those fish. And with platypus as well, they're in their burrows breeding, youngsters are coming out nowish [in late January], but you can imagine if you had a summer storm and they're in their burrows – somehow they know how to set the height of the burrow above that water, but then in that breeding period when they're in, December, January, that's not a good thing.'

I asked Sarah how she felt about the future of the river. 'I think the future's quite positive for the Yarra,' she replied, 'and a lot of our waterways. Melbourne Water co-designed [and] the community got involved [in] the healthy waterway strategy, so that gives us quite a clear vision – a combined vision – for each part of the Yarra and the rest of our operating area ... That document talks about the risks as well and mitigating for those risks, so I think there's a lot of knowledge out there, I think there's a lot of good work. A lot of friends groups do so much – I don't know where we'd be without friends groups. It's a privilege to walk on Country with Traditional Owners helping heal the Birrarung. Engaged landholders that are happy to have their waterway fenced out and willows removed and native vegetation replanted, that's a huge part of Melbourne Water's work. Because it's not just the flow in the river, it's many people putting together ... I think a lot of people love it and value it – personally I think the future's pretty positive.'

SHORT-FINNED EEL

Sarah Gaskill told me that I ought to write about short-finned eels if I was writing about the Yarra, and she's not wrong but the thing is I don't know anything about them. Which is not strictly speaking *literally* true because I know they migrate between the sea and the river but that's bare bones stuff; hyperbolically speaking (and only a little bit hyperbolically) that's nothing. For all intents and purposes. She did tell me that this switching between salt and fresh water makes them diadromous, which I think we can all agree is a very cool word, so that's fun to know.

I have an ongoing joke/debate with my girlfriend about swimming in the sea versus swimming in rivers. She's a sea person and I'm a river person (you might have already picked up on that by this stage of the book). Like most dichotomies it's a bit silly really because I like swimming in the sea with her and she likes swimming in rivers with me; at the end of the day water's water, right? But if you're a fish it's a bit more complicated than that. If you think back to high school science class you might remember learning about osmosis (another great word): it's the movement of water (or other stuff) from one side of a membrane to another. Fish absorb water into their body via osmosis – they basically drink through their skin, and as you'll be aware drinking water is pretty fundamentally important to the business of being alive. The thing is that you need to get the balance right: not too much water, and not too little. And the other thing is that the relative saltiness on both sides of the membrane (in this case, the fish's skin) has an effect on osmosis: if the water outside the fish is saltier than the water inside the fish, then water will be drawn out of the fish. And if the water outside the fish is *less* salty than the water inside the fish, then water will be drawn in. Which makes surviving in fresh water pretty challenging for fish that live in salt water, and vice versa.

So as you can imagine, being an eel is a bit trickier than just sometimes living in a sea and sometimes living in a river. Maybe now's a good time to do a run-down of just what an eel's life cycle is like (I'm learning all of this as I'm writing it, by the way, so it's as new and exciting to me as I hope it is to you). Here's how Victoria's Arthur Rylah Institute for Environmental

Research explains the life of a short-finned eel:

> Adults spend many, many years (decades) in freshwater habitats (such as rivers and wetlands) feeding and growing before metamorphosing into silver eels. The silver eels migrate downstream, often during high river flows, and into the sea to spawn. During these oceanic spawning migrations, adult eels travel several thousand kilometres to warm tropical waters. After spawning, the adults die and the newly hatched leaf shaped larvae (called leptocephali) commence a journey toward the coast, drifting on ocean currents and developing into glass eels before eventually entering rivers. The young eels, now known as elvers, migrate further upstream into freshwater, developing into yellow eels and eventually becoming adults. These adults will then return to the sea to spawn.

So count that: you've got 'normal' adult eels, silver eels, leptocephali (yet another great word!), glass eels, elvers and yellow eels. That's *six* distinct life stages just for one species of fish. All that and they travel thousands of kilometres over the course of their life, too.

It stands to reason that the short-finned eels you're most likely to see are adults, just because they can be around for decades. You might imagine that decades-old eels would be pretty big and indeed they are: nearly a metre long and as thick as your arm. I saw them once in one of the ornamental lakes in the Royal Botanic Gardens in Melbourne, near the city. I didn't realise at the time how special it was to see them; I've looked in the water every time since when I've been there but so far I haven't seen them again. In fact those lakes were once part of the Yarra, before that part of the river was straightened and they were cut off. I'm not sure if they're still connected underground like some of Melbourne's isolated urban waterways are, but even if they're not it wouldn't matter to the eels: apart from changing their body half a dozen times and migrating thousands of kilometres between salt and fresh water eels can also crawl over land if they have to. Because of course they can. Honestly at this point I wouldn't be surprised if it turned out they can fly.

A lot of people find eels off-putting or unsettling or scary or just plain ugly, I guess maybe because they don't look the way a fish is supposed to look: they're smooth and slimy-looking and snaky. I love snakes too so it's probably no surprise that I love eels. (You may be shocked, though, to learn that there is an animal I'm not so fond of. It's sloths; they look like demonically possessed muppets and they creep me out. But since they're not a species that is particularly relevant to a book about the Yarra we'll just leave it at that.)

I went camping in Budj Bim National Park in February 2023 and at the end of the trip I visited Tae Rak Aquaculture Centre. Budj Bim was World Heritage listed by UNESCO in 2019 because it's home to the oldest known example of aquaculture in the world. Short-finned eels have traditionally been an important food source for Gunditjmara people around Budj Bim, just as they have been for Wurundjeri people along the Yarra. At Tae Rak you can learn all about eels and you can also come face to face with them: there's a round tank outside the centre and you can lean over the edge and peer into the water; when you do a whole bunch of eels will come up and check you out. It's very endearing. The water catches the light from the sky and reflects it back to you at one angle while at another angle it shows you the dark green of the bottom of the tank, the two colours constantly jostling so that your mind becomes dazzled and your eyes struggle to make out anything other than the constantly moving surface – but then you suddenly see a wobbly tube of fish nosing up the water column towards you and then a snout breaks the surface, coming to a rounded point like a spearhead. Two yellow, tubular nostrils poke out from above the eel's slightly down-curved mouth, making it look just a tiny bit like a slug. On the sides of the head two large, flat, grey-blue eyes are just able to stare directly at you as you peer down (though apparently eels don't have very good eyesight). The eels twine around each other as they rise up to the surface of the water, their round pectoral fins fanning the water. They move slowly, deliberately, and it's all quite magical: it feels as though they're as curious about us as we are about them.

Every time I'm face to face with an animal I find it impossible to deny its individuality. What is a personality, in a wild animal? How long would I have had to spend staring into the eel tank before I started to discern some variation between individuals – this one a little more outgoing than that one, that one a little more watchful than the other one? What behaviours are instinctive and what are learned and is there a meaningful difference between the two? I haven't heard of anyone studying those particular questions in short-finned eels, but that doesn't mean that we haven't learned some extraordinary and fascinating things about them. Here's the Arthur Rylah Institute again:

> In 2018, we started a project using innovative technology to identify these eels' migration routes and spawning areas and

> describe the environmental conditions they experience as they migrate … Migrating adult eels were tracked for up to five months, and up to about 3000 km from where they were released. Tagged eels were tracked migrating to an area in the Coral Sea near New Caledonia, a presumed spawning area in the south Pacific Ocean. The eels moved up and down the water column; mainly between depths of 700–900 m during the day and 100–300 m during the night, likely related to avoiding predators and for thermoregulation.

Each question that science answers just creates a dozen new questions. That's how it works. You burrow down and down and down into the nitty-gritty of things only to find that it somehow gets even nittier and grittier. I'd love to know what an eel thinks as it moves up and down the water column, up and down the continental coast, as it grows and metamorphoses again and again and again, smells the salt of an ocean one year and the tannin of a river the next. How do all those impulses and sensations play across its mind? Does it anticipate the fresh water, or the salt water? Can it remember being a younger eel, in a different environment? Do some young eels watch and wait for other, bolder eels to begin their migration, and then join in? Do they find themselves pulled by something deep in their bodies, a line as invisible as an ocean current tugging them back to the river? Is it exciting, does it thrill them to be on the move? I can't help wondering about all of this. But in a funny way I love the not knowing, too.

DRAWING PEOPLE BACK TO THE RIVER

It was just after dark on 5 April 2022. The city lights were glinting red and blue and yellow on the black water of the Yarra, and Felicity Watson was trying to show me the Turning Basin, but Crown Casino was making it difficult. 'We know you're there, Crown,' Watson said, glaring across the river with equal parts irritation and amusement as a series of enormous decorative fireballs flared into the Melbourne night sky, roaring like dragons. 'You run this town!'

Felicity was in the midst of explaining the evolving history of Yarra Pools, the organisation she'd been president of since 2019. 'We were first incorporated as an organisation in 2014 and it was started by Mitra Anderson-Oliver and a group of passionate landscape architects, lawyers and urban policy advocates,' Felicity told me as the fireballs roared. 'The original spark of the idea came from the urban pools appearing across the world, inviting people into a closer relationship with their river systems and motivating massive clean-ups of waterways. In 2015, Yarra Pools merged with Yarra Swim Co, an entity created by designer and strategist Matt Stewart. Matt had brought back the historic three-mile swimming race from MacRobertson Bridge up to Princes Bridge.' Once a major event on the Melbourne calendar, the race had last been run in the 1980s. 'Matt was like "Why can't we do this again? We need to bring back the three-mile swimming race." So that started this odyssey that's evolved into something that's in a way a little bit different.'

Different indeed: instead of a race, Yarra Pools now envisages a more sedate enjoyment of the water. After regulatory hurdles made the revival of the three-mile race a non-starter, 'Matt thought about other ways that Yarra Pools could advocate for swimming to be reintroduced, and other catalyst projects, and that led to the idea of a pool in the river'. While a race might create havoc with boat traffic on the lower Yarra – it's not actually legal to swim below Gipps Street in Abbotsford – a pool is in a fixed location, and can be more easily avoided.

So what might a pool within a river look like?

Felicity picked up the story again: 'So Matt in about 2015 and '16 was working on purification technologies, so actually coming up with novel ways to filter water to make it swimmable, so he was doing some research on that and had some partners who were looking at developing swimming technologies to do that ... they looked at filtration technology, and then moved on to the idea of a floating swimming pool.' She summarised the decision-making process: '[There was] a question of "Do we want a pool that has river water, or do we want a pool that's next to the river that has just normal filtered water?" and that was a real challenge because filtering the river water is technically challenging, it also fluctuates in terms of pollution and all of that sort of stuff.' Perhaps being immersed alongside Yarra water would be enough, without actually being immersed *in* Yarra water. '[The Yarra Pools founders] felt that it was a piece of infrastructure that could have a legacy. It's not just for one day a year, it's going to be there to be a facility for the community to use for years to come. So I think the vision got bigger, because they were thinking of that. And a lot of people have said, "Well why didn't you try and do a pool further upstream, look at areas of the Yarra that are swimmable and where it might be less of a challenge in terms of engineering and everything?", but it was really important to us and to Yarra Pools to re-establish or to build on this connection in the CBD, where there's a huge population, there's all of the working population of Melbourne, and to reach the most people and to build a facility that could be of benefit to the most people.'

The idea of swimming in the river might seem alien to many contemporary Melburnians, but not to Felicity, who brought her passion for river swimming with her to Yarra Pools: 'I grew up in Grafton, in northern New South Wales, so I grew up swimming and kayaking on the Clarence River, and I loved it as well, and other places around the Northern Rivers. And that was just a huge part of my childhood, and so that's why this opportunity to get involved in Yarra Pools appealed so much to me – because it kind of connected me to this part of myself that I hadn't been able to experience for a really long time – to try and advocate for that to happen in this city is really amazing.'

And even if a pool in the Yarra seems strange, there's no shortage of precedents elsewhere in the world. 'I lived in Sydney for a long time,' Felicity told me. 'Everyone knows [Sydney's] harbour pools and the beach pools, and that kind of culture of outdoor swimming is really ingrained in the culture of Sydney and the harbour. But here in Melbourne it's not part of our culture in the same way. And some of it is probably to do with Melbourne weather and being a bit colder, but there are pools in northern European countries that are used all year round.'

With the idea of a pool decided on, the next thing needed was a location. Felicity came to Yarra Pools from working at the National Trust, 'so my professional background is in cultural heritage management ... and as part of that work we advocate for the protection of cultural, Indigenous [and] natural heritage, a broad range of heritage across Victoria, and so this project really appealed to me because projects like this are a really amazing opportunity to acknowledge history and interpret history as well, and particularly this location.' Felicity faced the river: 'The Turning Basin was the location of Melbourne's first port, it was the place where the *Enterprize* landed ashore in 1835, so it's a really heavily symbolic landscape of colonisation which is really important to acknowledge, and one of the things that we considered in our design, by WOWOWA Architecture, was how that history could be acknowledged, and how traditional owners could be brought in to support the interpretation of Aboriginal history and this site as a place of potential reconciliation.'

Felicity described the Turning Basin and adjacent Enterprize Park as 'not a well-loved part of the city, and we thought that this would be a really amazing way of activating it', but acknowledged that 'underused' is not the same as 'unused'. Enterprize Park has, at times, been the site of a well-established homeless encampment, and Felicity stressed that Yarra Pools is eager to avoid further disenfranchising the city's marginalised communities: 'That's exactly what we were trying to avoid, and I think that's the importance of civic-led projects in our communities, but they're also really, really challenging, because we're not a company, we don't make money, we're all volunteers.'

Regardless of whether Yarra Pools goes from dream to reality, the human landscape of the river is changing – in the city as elsewhere. During the course of our conversation Felicity brought up GoBoats: 'these little electric boats that have picnic tables in the middle'. The boats can be hired – 'anyone can hire it, you don't have to have a boat licence' – and taken out joyriding on the river: 'You can zoom right down to the Bolte Bridge and up to Herring Island and it's getting people on the river in a way they probably didn't before,' Felicity said, a warm glow entering her voice as she also brought up various riverside bars that have sprung up in recent years. But then she demurred: 'I have mixed feelings about it because the river's a public space, it's actually blue space in the same way that green space is public space, so when you commercialise bits of it, it kind of eats away at its integrity, but the flipside is it gets people there.'

As lockdown brought people into their local parks and gardens, so it brought them to the riverside – and especially to see Salvatore, Melbourne's celebrity seal of 2021 (more on him in the next chapter). I asked Felicity if animals would be allowed in the pool – if the Yarra Pool would become like the coastal pools of Sydney, which have their own ecosystems that develop as sea water crashes over the concrete perimeters. 'Well I wouldn't be against it,' Felicity said. 'It's like the whole debate about do you try to incorporate the natural environment, or does it have to be a separate, sterile environment, and one of the issues is that ... the threshold for water quality under our current laws is super high, and it's arguably higher than it needs to be to be safe, and so there's this really high threshold for being able to swim in waterways. [It] is obviously important to be able to regulate that and keep people safe and to address the sources of pollution and things like that, but it makes it a really high bar to overcome and that's why we ended up going with a proposal that was next to the river but didn't include actual river water. But we even thought about doing things like making the water look like the colour of the Yarra River, like a browny colour and not the blue colour that you'd normally expect to see in a swimming pool, to make that mental connection.'

Despite setbacks and changes of plan Felicity sounded

hopeful about the future of the river: 'I feel like I've dropped into this campaign like a shepherd to keep it going ... but I feel like it's going to continue regardless, and it's building momentum, and it's something that's really important for our city to connect to, and a swimming project is a really great beacon for what's possible in drawing people back to the river. So we're going to be looking at ideas like floating saunas and little pop-up pools and things like that. So it's never going to die.'

Only a few days before my conversation with Felicity, Melbourne Lord Mayor Sally Capp had tweeted an April Fools' Day joke about city workers swimming in the river. After that 'we thought we should invite her to jump in with us', Felicity joked. I asked her if she'd jump in, too: 'Oh yeah, I'll do it,' she replied, without hesitation. Then I asked her if she'd swum in the Yarra herself and she replied with mortification: 'No! Oh my god, no one's ever asked me that before!' Then she offered an explanation: 'I know this is absolutely no excuse, but I was elected president at the end of 2019 [and] almost directly after that I broke my leg so I was in a cast for six weeks. And then COVID happened. So it's been weird to be the leader of this project during this time because most of my colleagues on the Yarra Pools board I haven't met in real life, we've just met over

Zoom, and when you're trying to plan a project that's very place-based and community focused, that kind of feels weird, but it is what it is. But I will take up that challenge.'

Talking to Felicity, it was hard not to be drawn in by her enthusiasm for challenges – and for the river. Since our conversation, I've had in my head a joyous vision: of leaving my day-job office in the city at the end of a working day, the sun still high and glinting off the glass towers of Melbourne's CBD, and catching a tram down to Enterprize Park, and changing into my bathers and leaping into the water – perhaps meeting friends there, surrounded by other city workers fresh out of their air-conditioned offices and splashing in the water on a hot December evening. Why, I want to know, can't city life be like that? Why shouldn't it be?

AUSTRALIAN FUR SEAL

Once, maybe a decade ago, a couple of friends and I made a dash from Melbourne down the Bellarine Peninsula to Queenscliff because somebody had seen orcas there. Of all the animals in the world that I long to see in the wild, orcas are probably number one (it's a particularly marine list: rounding out the top three are octopuses and sharks – any kind of octopus, any kind of shark, I'm not fussed). Of course by the time my friends and I got down there the orcas had long gone: I don't know if they can swim in the ocean more quickly than a person can drive a car, but it certainly comes more naturally to them. But Queenscliff is a lovely little town and orcas or not we had a lovely little time, spent the afternoon there, had a drink at the old hotel, went to see the lighthouse. (Several years later I was back at that lighthouse, after taking part in the Queenscliffe Literary Festival, and a storm was blowing in and I was watching albatrosses dancing in and out through the Rip.

A certain heralded and garlanded luminary of Melbourne and Australian writing was there, too, and had she asked me what I was looking at I would have gladly shown her some of the most spectacular birds on earth but she didn't, and obviously I'm biased but I'm inclined to think that it was her loss.)

Anyway, the point is, my friends and I didn't see any orcas, but we did get to see seals frolicking in the surf below the lighthouse. And you can't have a bad day if you see a seal; seals make everything better (unless you're a fish).

There are quite a few seals in Port Phillip Bay, and surrounding it. They're all along the Victorian coast, actually, lolling about like dogs, poking their heads up out of the water to gaze at passing boats, lumbering over rocks. It was only when I saw them at Melbourne Zoo that I really understood them: in the seal display there you can go below the level of the water and look through the glass and see them zipping around, literally in their element. On dry land they look clumsy because they're not built for it. Even an Olympic swimmer would look clumsy to a seal if they happened to encounter each other in the sea (or in the pool). That's okay, you can't be the best at everything. We have our habitat and seals have theirs, and in their proper habitat seals are just about unbeatable, as if someone had sculpted water into flesh. (Smelly flesh, but flesh all the same. I once went to Cape Bridgewater, in the far southwest of Victoria, where you can walk along the tops of the cliffs and see the state's largest seal colony a hundred metres down below, and smell it too. Let me just say that it's a good thing those baby seals are cute because standing above their home was like standing above an open sewer. Incidentally, someone had seen a great white there the day before, and I'm still upset that I missed it.)

I guess maybe we tend to associate seals with seal colonies, but not every colony is a breeding colony: sometimes they'll just be haul-outs, places where single, young seals like to drag themselves up out of the water and hang out. And sometimes a single, young seal, wanting perhaps to see something of the world, will go on a bit of an adventure, and swim wherever the water takes him (because it's the hims that are particularly prone to wandering), and that's how you end up with a seal in the Yarra.

It seems to happen every few years or so. There'll be excited videos posted on social media, and local news outlets will pick up on it, and away we go. It happens sometimes with dolphins too, but they don't seem to be as prone to venturing up the river. The seals who turn up don't usually hang around for long: a few days, and then they're back out. Once, when my day job had its office in Docklands, and before the Docklands skyline got filled up with other office blocks and apartments and our view disappeared, I arrived at work to find several of my colleagues out on the balcony, looking at something: it was that year's seal, swimming through Victoria Harbour en route to the mouth of the Yarra, under the Bolte Bridge and back out into the Bay, back home.

In 2021, though, one seal hung around long enough to become a celebrity, and hundreds of people saw him.

When a seal enters the Yarra it has a defined zone: Dights Falls, in the shadow of the Eastern Freeway next to where Merri Creek flows into the river, forms a natural barrier to the pinniped's peregrinations. That still leaves a lot of river frontage to explore, though: a dozen kilometres or so of winding, bending Yarra. It's the part of the Yarra that people are banned from swimming in because of boat traffic, but try telling that to a seal. In fact it was probably because of a lack of boat traffic that the 2021 seal hung around for so long: the lower Yarra was unusually quiet throughout much of that year as lockdowns restricted most boat traffic on the river below Gipps Street.

Because people like to give names to wild animals whom they take to heart, an earlier seal had been named Salvatore: he liked to swim, at least so it was said, behind the Salvos Store in

Abbotsford. And because people like to imagine that beloved animals will return to see them again, the 2021 seal was also named Salvatore. Here, though, I must play the party pooper a little. (Well actually I mustn't, there's really no need to, but I'm going to because I find these facts genuinely interesting and also, I'll be very blunt with you, because I have a word count to meet.)

You see, Salvatore, the swimming-behind-the-Salvos Salvatore, he wasn't just a Yarra seal. First of all he was a Maribyrnong seal; he was seen there beneath a pier, tangled in packing tape. He had to be cut free, if I remember the story correctly, but the tape left scars – telltale scars. And as we all know, scars don't disappear.

And you see, that lockdown Salvatore, the great Yarra seal of 2021 – he didn't have any scars. You can check for yourself: there's no shortage of photos and videos of him; hundreds if not thousands of Melburnians must have seen him and the vast majority of them would have been carrying phones, and a phone camera that goes unused in the presence of a surprise seal is a phone camera wasted, at least as far as most people are probably concerned (certainly me). I've seen a lot of footage of Salvatore the Second and I haven't seen any evidence of scars. And when I saw him myself, in the flesh (I'll tell you about that in a minute), as far as I could tell he was unmarked. (He was also magnificent, not that the two things go hand in hand – or flipper in flipper.)

But it really shouldn't surprise us that more than one seal has seen fit to swim up the Yarra. For starters, there are a lot of seals out there in the coastal waters of Victoria. And like carnivores all over the world they're curious animals. Secondly, the Yarra is a great resource if you're a seal: all those fish, including lots of big, fleshy carp, and no other seals to share them with, and the fish have never had to deal with anything bigger than a cormorant before. For a fish happily lounging about in the Yarra the sudden arrival of a 2-metre-long, 300-kilogram Australian fur seal must be like Godzilla rolling into town. A very large proportion of all those videos and photos of Lockdown Salvatore showed him slapping and smashing carp on the surface of the river, swinging them with their tails held between his sharp clenched teeth – and looking like he was having a whale of a time while he was at it. (I wonder: can a seal have a whale of a time?)

Whatever his provenance, though, Salvatore the seal of 2021 was a star. It might be a little too cute to call Salvatore the city's salvation but it wouldn't necessarily be inaccurate, either. He was at least a salvation for those of us who were lucky enough to live close to his stretch of the river, or who are unlucky enough to live our lives glued to social media (guilty, both counts). Whenever there was a lockdown in Melbourne in the first two years of the pandemic and people were restricted to a radius of only 5 kilometres from their house, the same phenomenon occurred: every park, every green space, every pleasant pathway through natural or near-natural surrounds became busier than the steps of Flinders Street Station at peak hour in any year BC (Before COVID). This was certainly the case for the Main Yarra Trail, very much including the entirety of the dozen or so kilometres between Dights Falls and the city that was Salvatore's domain. It was hard to find some peace and quiet, harder still to find any social distancing. Everyone was on edge even if they didn't realise they were. Joggers and cyclists whizzed and sometimes wheezed past people who were taking a more leisurely approach to their exercise, everyone squeezed onto one fairly narrow bike path with traffic going both ways. In the weeks of Salvatore I was usually on my bike, riding as carefully through the crowds as I could, because that allowed me to cover more ground more quickly and thus be more responsive to any sightings of our pinniped hero (pinniped just means seal; I'm showing off).

And because I was on my bike, I passed more people than I would have if I were on foot, and because I passed a lot of people I overheard a lot of conversations, and a *lot* of conversations in those weeks were about Salvatore.

The conversations were always about 'him', as if he was some kind of river god – which he may as well have been given all the wonder and awe he inspired: 'Have you seen him?'; 'I saw him yesterday'; 'He was right here just a few minutes ago, really!' The best conversations to overhear – only ever in fragments, only ever in passing – were those in which the listener was previously unaware of the seal, and the teller had the joyous task of imparting this wondrous new knowledge to them. A seal in the Yarra? Truly? Yes, on my word!

There was a lot of concern for the seal and his welfare, both along the riverbanks and on social media. People worried if he was okay, if he was lost, if the river was a health hazard for him. But the abundant footage of him swimming up and down, appearing everywhere from the city to Abbotsford, and smashing hapless fish senseless on the river's surface suggested that he was doing just fine, thanks.

I saw him twice, though I tried to see him many more times than that. I'd type the magical phrase 'yarra seal' into Twitter's search bar and refresh the results every few minutes, obsessively, and I had my bike helmet and binoculars ready to go so that I could up and leave at a moment's notice if he was seen nearby – and given that Dights Falls, one of the most popular spots to see him and the upper extent of his range, is only a five-minute ride from my house, he was seen nearby quite a lot. I missed him by minutes on more than one occasion. I'd see someone with a camera staking out the riverbank and ask them if they'd seen the seal and they'd tell me that he'd been there just a few minutes ago, and they'd last seen him going upstream behind the Children's Farm, and I'd dutifully ride upstream but he'd never appear. There are certain blind spots in the river where a seal can turn around and change its course and nobody can see it happen, or at least nobody riding a bike on the Main Yarra Trail.

I started getting off my bike and walking the rough tracks on the other side of the river, tracks I'd walked many times before, now newly smoothed by the hundreds of people tramping up and down them every day in those long lockdown weeks. For better and for worse many of the walking tracks now evident along the river were made more navigable in those weeks. (Some, unfortunately, were also opened up, by people desperate to escape the throngs and other people keen to see where these new tracks went – I admit I was not always immune to such following urges myself, though I regret it now.) Where the Main Yarra Trail passes the Children's Farm the riverbank is fenced off, not visible from the sealed path, but following the walking track on the opposite bank you can see the whole thing, and I walked that short stretch – no more than ten or fifteen minutes from end to end – back and forth, back and forth, looking for the seal and familiarising myself, too, with the geography of that unfamiliar part of the river: the muddy unnatural beaches where, perhaps, a seal might like to haul up (after all, he must sleep somewhere); the fallen trees, half submerged, where he might like to hunt (fish like to hide under submerged debris); the sheltered pools amid the eddies and currents where he might like to bask.

But when I saw him first he was behind the Carlton and United Brewery, where invasive wandering trad forms dense thickets down to the water's edge and equally invasive willows overhang the water, and where the toasty smell of malted barley sometimes infuses the air.

I'd got the alert from a couple of friends, messages on various platforms pinging on my phone. I jumped on my bike and raced off for the river. I'd been stung too many times by the seal's slippery ways, how he'd double back and disappear, so I tried to be clever – as if I could outsmart a seal! At the Gipps Street bridge in Abbotsford, next to the Salvos Store that had given this and previous and probably every future seal to swim in the Yarra its name, I was faced with a choice: I could cross the bridge into Andrews Reserve, from which I knew I could continue upriver for a couple of hundred metres along a path that offers fleeting and erratic views of the river and soon becomes too narrow and winding to ride on, at which point I'd be stuck with my choice whether right or wrong; or I could stay on the left bank of the river, and ride around the brewery to a small park just upriver where I might get a panoramic view both up and downstream. I'd never been to that small park before. I knew about it only from looking at Google Maps. I didn't know if it would offer me what I wanted, but it *looked* like it *might*. I decided against sticking with what I knew. It's not like me, but I panicked, so desperate was I to see the seal after so many failed attempts.

The park did not offer the views I hoped for. I suppose the river just winds too much. I could see as far as the next bend in either direction, but that wasn't very far. By the time I got there and realised this I'd wasted precious minutes: ten, fifteen of them. I was dripping with sweat and I had no way of knowing how long the seal would stick around for. My friends were still sending me excited updates from the other side of the river, from some stretch I couldn't quite see from my location. To get there, to ride all around the gigantic brewery again, cross the bridge, ride to where they were, would take another ten or fifteen minutes. But what had I to lose? I jumped back on my bike, heart racing.

I found my friends crouched amid the fallen wattles and red gum saplings high above the river, overlooking the humming concrete of the monolithic brewery just across the water. They told me how the seal had been there minutes earlier – a tale I'd heard so many times before. I braced myself for disappointment. But then – oh! With an exhalation like a whale surfacing from the depths, he appeared. Dark-brown back glistening with river water, arching in a slow half-moon above the surface like the Loch Ness Monster before disappearing again, making gentle but forceful motion upstream against the river's flow. And again! That same sound: a hissing snort of air exhaled under pressure and at high speed; an instant's silence of undetectable inhalation, then under again. How many times did we see him? Over how many seconds? A mere handful? We rushed to the Gipps Street bridge to intercept him, but he never reappeared. At some point he'd turned around, or had accelerated and beaten us; perhaps he'd opened a rift in space and swam serenely through it to somewhere unknowable – he may as well have. Like the platypuses two dozen kilometres upstream, like the cormorants and coots and darters all along the river's length, once he dived beneath the brown, clay-thickened water he was gone, his reappearance just a guess and a hope.

You'd think that sighting would satisfy me, so long had it been in coming. But it just made me want more. That mere taste was not enough. I redoubled my efforts, leaned ever harder into my obsession; it felt as though the entire city was undergoing the same experience. All the city were seal watchers, or watchers of the watchers, on social media, in mainstream media – there were those who lived within 5 kilometres of the seal's domain, and those who wished they did. How fortunate I felt to be in the former group.

The days were long and sunny and I'd take extended lunch breaks from work, or take no lunch breaks at all or get up early so as to finish my day as early as possible and go looking. I longed for weekends to come so that I could spend my hours – as many as lockdown allowed – along the river's edge, striding back and forth, back and forth. I revelled in the wattles dropping their blossom into the slow water, the slick of festive gold as pollen washed downstream on the currents. Small-leaved clematis flowered, releasing its sweet creamy scent into the air; tree violet followed, its uncountable legions of tiny flowers wafting their heavenly smell into the air in such a way that it caught passers-by at odd moments, when they least expected and most delighted in it. Sticky boobialla began flowering, and I learned its name at last; this pretty shrub I'd walked past so many times, now a regular companion beside the rocky tracks along the river opposite the Main Yarra Trail. So many other people were walking the same tracks, all in their own worlds, many also looking for the seal – although some, to my enormous surprise, hadn't heard of the seal, as I found out when I'd ask them, smiling with enthusiasm: 'Have you seen the seal?' And then I'd tell them all about the seal, and how extraordinary he was, and I'd doubtless appear as loosened as any other balding, single forty-something man talking enthusiastically and slightly unintelligibly about something that excited his passions. On the whole I was treated with commendable tolerance, I think.

And then one day it finally happened. My luck changed – or perhaps the seal decided to finally grace me for all my efforts. It started on the Main Yarra Trail, just before Gipps Street, with a group of people clustered on the grassy bank and me pulling my bike to a stop just in time to see the seal, the entire enormity of him, haul briefly out of the water on his way upstream to the Children's Farm. Again that sharp hiss of exhalation, a sound I'd been listening for up and down the river since I first heard it. The seal arcing three or four times out of the water. And me with a decision to make once again, a gamble, taken with uncharacteristic decisiveness: roughly calculating the seal's speed I jumped back on my bike and raced to the Johnston Street Bridge, up the hill and over it with the river far below, locked my bike up and made my way to the dirt track, where I began walking briskly downstream, my intent to intercept the seal as he – hopefully – continued swimming sedately upstream. It was a gambit I'd tried before but this time, I was sure, I had the timing right.

I made it to the bank opposite the Children's Farm, and there was no sight of him. There are continuous views of the river along this track and I could neither hear nor see a thing. 'Has the seal come past?' I asked mystified picnickers, to which I received blank stares, confused shrugs. I kept walking at a brisk pace. I rounded a bend in the river that took me beyond sight of Johnston Street; I scanned the muddy beaches on the opposite bank, those small coves hidden behind the Children's Farm. No seal. I kept rounding the bend till I could see the bike path on the other side again, beyond the furthest edge of the Children's Farm, and I could see also the crowd of people I had, only ten minutes before, been among: they were still watching, and pointing, and as they did so I saw the seal rise again: he'd turned around once more, or halted his progress, and had barely moved from where I left him – but as I watched he resumed his procession upstream, and now he was coming towards me.

Before I knew it he was opposite me, on the far side of the bank: he spent some time investigating a fallen tree, searching perhaps for fish sheltering beneath the submerged branches, as I'd surmised he might. When he swam away from the tree he arced up out of the water like a whale breaching and let his full, mighty weight slap back down onto the surface of the river and through, creating an enormous bow wave that slapped against the riverbank again and again in its reverberations, inundating the bank. He dived, and sometimes disappeared for what seemed like minutes before resurfacing further upstream, or further downstream; in these moments people on canoes would occasionally paddle by and I shouted out a warning to them: be careful, there was a seal here just a minute ago and he's *massive*.

Because he was. I hadn't appreciated it the first time I'd seen him, behind the brewery, and the hundreds of photos and videos I'd seen of him certainly didn't convey it, but he was a huge animal: easily 2 metres long, seemingly built of pure muscle and power. No wonder the gentle flow of the lower Yarra couldn't hold him back; no wonder he scoffed at the many raging floods of that flooded year. When he rose from the depths he was like a leviathan. I was very glad I wasn't a fish.

Very often, I saw, he rose upside down: his chin and tongue would surface first, and then his nose and his whiskers splayed like a cat's, and he'd breathe out in that now-familiar way and then dive again backwards so that the top of his head went under first. At first I worried that he was ill, discombobulated by some contaminant in the water he was immersed in – but then I recalled how vibrant and strong he appeared otherwise, how agile and active. Then I thought, and I thought this for several days, that perhaps he was swimming upside down so as to catch his meals by seeing them from below: perhaps, I thought, in the murk of the river's water it was easier to swim along the bottom upside down, and look up towards the surface, and so see any fish above silhouetted by the sunlight. But then a friend told me that she'd heard that seals hunt by feel, using their long, sensitive whiskers to scan the water and detect by touch anything that might be worth eating. So to this day I have no idea why the Yarra seal was swimming upside down, and the only thing I can think is this: that we're fools if we think we humans are the only animals in the world to ever do something just for the hell of it, and that the Yarra seal seemed to be living a pretty good life, with a river full of fish all to himself and none of them wise about giant predatory mammals, and that maybe life was sweet in those long weeks he spent in the river, and all in all he could afford to muck around a bit, and who am I to say that a wild seal can't sometimes be a bit of a goof when the sun's shining and the fish are jumping.

Shortly after I saw the seal behind the Children's Farm, where I watched him for twenty minutes, lockdown was eased and boat traffic returned to the mouth of the river, and the seal was gone again, as abruptly as he'd appeared. But there will be more seals in the Yarra in the future – in fact, there already have been.

THE WISDOM OF THE GROUP

'We actually started on exactly the same day,' Meg Elkins told me. 'Fran and I didn't know each other, but we both launched on the same day.'

'Yep,' Fran confirmed.

'The 29th of August 2021.'

'So in lockdown,' Fran explained.

We were sitting, all three of us, on the large, flat rocks scattered about Deep Rock, in Yarra Bend Park, a few hundred metres upstream from Dights Falls. It was January 2023, just a few days into the new year. Only a couple of months earlier these rocks had been under water; an information sign a few metres away, explaining the recent history of this place, had also been up to its neck in water. Now it was early on a Saturday morning and it hadn't rained in a couple of weeks; the morning sun sparkled on the river and joggers and dog walkers hurried past or stopped to linger.

The rocks at this spot form deep steps down into the river; below the surface algae grows on them, and on warm evenings – or any time you want – you can sit on them and dangle your legs in the water. It's a good way to build up to jumping in for a swim.

As I was sitting with Fran and Meg, Sarah Tomasetti came by in her swimming costume. They all greeted each other like old friends. 'Have you been in yet?' Sarah asked Meg.

'Yeah, it is absolutely beautiful today,' Meg replied. 'The sunshine on that bit around the corner, and there's very little – I was saying to Fran, this is the least amount of current I've had in about four months.'

'Have you been in, Fran?' Sarah asked.

'Yeah we swam together,' Fran said. 'It was beautiful.'

The river was moving slowly, a broad spread of bush debris floating on its surface: leaves, fragments of eucalypt blossom. On the far side, below the high cliffs that give this area such a striking aspect – and from which, the information sign explains, in 1918 Alick Wickham jumped a then-world-record height of 62.7 metres from a specially constructed platform into the river – occasional sticks floated in the gentle current, giving the appearance, if you didn't look too closely, of snakes, which sometimes swim across the river, minding their own business as they wind their way from one bank to the other.

Fran and Meg explained that before their first swim at Deep Rock, neither of them had experience swimming in cold water. When Fran first swam here she parked at Dights Falls, a ten-minute walk away. After her swim, on the way back to her car, she started to get the shivers – 'nearly hypothermia'. Meg had the same experience. 'But you rely on the wisdom of the group,' she said. In the early days the swimming group built a collective knowledge of how to handle swimming in cold water and cold weather: bring a change of clothes; have hot tea to drink afterwards. 'We came back, and we kept coming back,' Meg explained. 'The nature experience is beautiful, but the community is really strong.'

As we talked, Sarah met another friend – Carrie – and they both plunged in. Cicadas were singing all around us, starting up their day-long recitals. Sound carries across water and we could hear Sarah and Carrie chatting as they disappeared upstream around the corner. There's a buoy in the middle of the river there, almost out of sight, and the Yarra Yabbies – as the group of mostly women who've taken up swimming at Deep Rock every day call themselves – like to swim there and back. 'I swam to the buoy and back twice, my first time,' Fran said – acknowledging that it had been too long to stay in the water in the cold conditions. 'Because you swim breaststroke,' Meg explained, 'you can talk to each other.' Conversation flowed more easily on the river, she said: swimming side by side instead of looking directly at each other allows two or more people to speak more directly, less self-consciously. Meg described swimming to the buoy or back and catching a glimmer of conversation, and swimming across to join it – as if at a cocktail party. The conversations encourage people to stay in the river for longer, too, Meg reckoned.

I first got wind of the Yarra Yabbies from a Facebook post on my local community group in early August 2021: someone had seen people swimming at Deep Rock and wanted to know how to join them. The swims had begun in the middle of that year, the second year of the COVID-19 pandemic, out of necessity. The original swimmers, unable in Melbourne's rolling lockdowns to visit their usual swimming place in Port Phillip Bay, began instead swimming at Deep Rock, defying the conventional wisdom that the lower Yarra is too polluted to swim in safely. They started a WhatsApp group to keep in touch with each other – a group that continues to this day. Every morning and every evening my phone buzzes with messages from people in the group, organising the next morning's swim, asking if anyone wants to head down for a swim, checking if anyone's planning on swimming that evening.

'Early on,' Sarah explained after she and Carrie had returned from their swim that January Saturday morning, 'it was anyone could join the WhatsApp, and then they had trouble with – I say "they" because there was a core group who were dealing with it – they had some guy who was kind of creepy and made comments and stuff, so then they shifted it to "Okay, to join the WhatsApp group you need to talk to one of these six people" or ten or whatever it was.' While that might make the group sound exclusive, or guarded, it's anything but. When one of those core members, Carolyn Tate, added me to the group I had some trepidation about announcing my purpose: being a writer can feel a little bit like being a cuckoo in the nest at times, and I didn't want the group to feel I was there to exploit them for their stories. But I needn't have worried: as soon as I introduced myself, and explained what I was doing, putting the message out into the sometimes hostile anonymity of the internet, a flood of messages came back, the group welcoming me – as I've since seen them welcome others – and offering to talk.

It reflected what I'd seen in person, a year earlier: having got wind of this swimming collective I'd ridden down to Deep Rock from my home nearby a few times and had built up my courage to swim. I only learned later that the main daily swim happens first thing in the morning, but even so on the three or four occasions I went down there on hot evenings in early 2021 there was a communal warmth and kindness palpable around the flat-topped rocks. People arrived with towels wrapped around their waists, removed them to reveal the swimming costumes underneath and slid into the river as gently as seals sliding off an ice floe. Dogs leaped into the river all around them – some of them belonging to the swimmers, some not. People sat on the grass nearby, or at the picnic table, eating and drinking in the long twilight. It was an idyllic scene, repeated and repeated.

'It's been such an inclusive community thing,' Sarah told me a year after my first swims there, 'and I feel like people have been able to join and participate and generate things as much as they might have room for or need to.' Sarah herself is an artist, and had recently finished a PhD – incorporating her thoughts on the river and on swimming into her final thesis. She started to explain the gatherings that sometimes happen in the evenings during the warmer months and then trailed off. 'Actually there was a bench and table ...' she said, looking at the now-empty space where they had been.

'That got washed away in the floods!' Carrie exclaimed. 'Carolyn, I think today or yesterday, has actually organised for it to be dismantled and she's going to have it put back here and reassembled.' After searching high and low, Carolyn had found the picnic table washed up below Dights Falls, nearly a kilometre downstream.

Carrie's first swim at Deep Rock was in early July 2022 – 'a few days before the twelve-month anniversary of the [Yarra Yabbies'] first swim'. Before she started swimming, she'd been riding her bike to stay active: 'During lockdown when there was the five-kay limit I was doing a lot of bike riding around the rivers and creeks because that's where the paths are, and I could get in a twenty-kay ride within five kays of riding around the Yarra,' she explained. But she needed something more. 'It was the day after the *Guardian* article,' she recalled, referring to an article about the Yarra Yabbies. 'I had been saying, "I need to do something, I'm really stuck personally," and I kept thinking, "If I could swim in the Bay I would, but it's too far away. It has to be easy." And then ... I read that article and I came down straight away and I sat here and waited till people turned up because it was a Sunday, and then that afternoon I bought some booties and I came back the next day. So that was my first swim.'

'And when you say "stuck"', Sarah prompted her gently, 'and you needed to do something, that's interesting. Does the river answer that in a way other than just exercise?'

'No, not just exercise,' Carrie reflected. 'I've always exercised ... I needed to do something – how do I say it? It's really hard to explain but I needed to do something that would give me some love, I think. And the fact it was freezing – it was only about seven and a half degrees in the water on my first day ... I needed to do something ...'

'Something a bit jolting,' I prompted her.

'A jolt,' she agreed. 'Something substantial.'

'And something communal?' Sarah asked, sitting next to Carrie on the rock – just as half an hour earlier, as Meg had pointed out, they'd been swimming side by side, talking. 'Was that part of it?'

'That was part of it,' Carrie said. 'That certainly appealed. And I love swimming. I'll often look at the water and think, "Oh that looks beautiful but it's far too cold," and I now know that it's probably not too cold wherever I go. I've just come back from New Zealand and I swam in glacial lakes and rivers and it's not too cold ... So that's how I started.'

For her part, Sarah had been walking with a friend before she started swimming at Deep Rock. 'We got into a habit of walking every morning in lockdown and connecting to this area,' she told me. 'So we also would sometimes walk for hours, and went down paths we'd never been down before. So we explored this whole area and the other side of the river from the Studley Park Boathouse to Fairfield and over there. There's all sorts of trails everywhere. So we saw the swimmers a few times, and then I spoke to them ... I think when I went in it was around about the eleven, twelve degrees [mark], so it had come out of that really freezing winter.' She estimated her first swim at Deep Rock was in August or September 2021, continuing, 'I've never done it super-regularly like every single morning, mainly because I'm a crappy sleeper ... [But] I guess I've always loved swimming, I've always dreamed of being able to swim here because I live here, so it was a revelation.' She explained that one of the swimmers owns a gallery nearby, 'and he's moved it to a new [building] and he decided the last exhibition should be a show about Birrarung, and I helped him curate that and so we kind of opened it up to anyone who'd been engaged. Because people ... were writing about it, all sorts of things came out of the swimming that were interesting – and he decided to have it as a community-based show ... and [we] just invited everyone who had a kind of connection to it and it was nice to do something that was community-based and not necessarily based on the status of the artists or any of that.'

Near the start of her book *Yarra: A Diverting History*,

Kristin Otto lists the creeks and rivers whose waters ultimately flow into the Yarra. The list runs to four pages, each page containing two columns, each column containing the names of dozens and dozens of waterways. 'Melbourne,' Otto notes at the end of the list, 'is blessed with creeks, and rivers.' So why do we treat them so badly?

On my phone I have an app: Snap Send Solve. It allows me to take a photo of something – pollution, a fallen tree, a dead animal – and report it to the appropriate council or government authority. By far the most common report I submit is for pollution in Merri Creek – one of those Yarra tributaries, which drains the northern suburbs of Melbourne. It's a popular recreation route: the Merri Creek trail runs from the Western Ring Road right down to the Yarra, a few hundred metres downstream from Deep Rock. On any given day hundreds of cyclists and pedestrians use it, to get from A to B or just to get out of the house. But swimming in the creek is unthinkable.

So too for any of Melbourne's other creeks or rivers that you may care to mention. True, people may swim here and there, but when it comes to leisure time in the water this city of five million people looks either to public pools or to the Bay. According to Melbourne Water there are 8400 kilometres of rivers and creeks in greater Melbourne – but you can double that, because each river and each creek has two banks. By contrast the shoreline of Port Phillip Bay is 264 kilometres long – only 22 kilometres longer than the Yarra River. But it's assumed that you can and should swim in the Bay – whereas it's more normalised by far to hear about dumping in Melbourne's rivers and creeks than to hear about people swimming in them.

'All the pools were shut because of COVID,' Sarah told me, recounting the Yarra Yabbies' origins. 'So I feel like COVID went like this [she made a noise like a balloon rapidly inflating], the lockdowns and everything, and the swimming group and the connections that came out of it went [now a noise like something darting out of the way very quickly], and people were like "Oh my god! Oh we can just get out of here". It was like this little escape ... There's no entry fee and you can safely engage and it's kind of magical.' Australia prides itself on being a nation of swimmers, so if Melburnians are going to swim why should it only be those people who are able to access the Bay, or who can afford to swim in a pool?

'I think it's important not to shy away from the fact that it is polluted,' Carrie warned, 'and it needs our care. And I think there's a great sense in the group that we don't want to just take from the river, which is where the clean-ups are [playing a part] and people writing about it that want to advocate for the river.'

'And invest in it,' Sarah added. 'Invest in it so that there's a care and drive towards policy or changing things in the ways that they need to be changed. I think there's been an improvement ... but there's still more that can be cleaned up. And I feel like the more people that are involved in enjoying it and wanting to rehabilitate it, the more chance there is.'

'Being in the river,' Carrie said, 'I've never felt so much like I am in the environment ... Even on a bushwalk, I'm walking through it, I'm looking at views, I'm looking at plants, but I found when I got in the river, particularly when it's so cold, once you're in that river nothing else exists when it's eight degrees.'

It wasn't eight degrees when I first swam in the river, at Deep Rock a year before I met Carrie and Sarah and Fran and Meg. I started slowly, uncertainly: on a hot night early in the new year of 2022 I sat on the rocks at Deep Rock and took my shoes and socks off and dangled my legs in the water, up to my calves, watching as my feet disappeared beneath the surface. The river washed gently against my skin, and softly.

The next time I went back, the next night, I felt braver. I wore my swimmers, and I stripped down to them and inched into the water. Even on a hot night the water was cold enough for me to be tentative – I'm terribly slow to immerse myself in water; I have to force myself to do it if it's anything below twenty degrees, though I always love it while I'm in. When I stepped off the rocks I felt beneath my feet a fine layer of silt: the same fine clay particles that give the river its colour; probably algae, too. It took some getting used to. I walked until I was up to my shoulders. A dragonfly buzzed past my head. Out in the middle of the river I could see the current washing a line of debris, but at the edge there was nothing on the water. I stayed on the edge: I'm not a strong swimmer, and I didn't trust myself to be able to swim across to the other side. I swam back and forth along the riverbank a few times, breaststroke. Mostly I just enjoyed being in the water, up to my neck in it.

It was a cold year, 2022, a La Niña year, so I only swam once more after that. I swam in other rivers, elsewhere: Canberra; Howqua Hills. I dipped my toes in a creek on the Alpine Crossing in Victoria's high country and it was so cold that it was hard to believe it wasn't frozen solid – that's as far as I got with that one. I splashed the water of the Plenty River over my head on a hot, sweaty walk through Plenty Gorge. Then the new year of 2023 rolled around and I turned my swimming thoughts back towards the Yarra. On New Year's Day I swam with my girlfriend at Laughing Waters in Eltham, where a small crowd of people had gathered. The bank was badly eroded and somebody had tied a rope to a tree so that people could lower themselves down to the water's edge. It was beautiful in the water but I'm not sure I could swim there again: if the riverbank isn't being cared for, then the river isn't being cared for. I have, at times, placed my own wants above the river's needs. I'm neither alone in that nor immune to it. I'm trying to be better.

Then I returned to Deep Rock. The large slabs of rock there, cut into cube-like shapes and placed as stepping stones presumably by the Deep Rock Swimming and Lifesaving Club in the first half of the 20th century, seem a kinder way to enter the river. After talking to Carrie, Sarah, Fran and Meg I went for a swim – or a stand. There's a river red gum, perhaps a few decades old, growing right next to the river at Deep Rock and if you inch along the river bottom by the rocks you can find one of its roots, protruding above the silt. In deeper water it could be a dangerous snag but in the shallows of the river's edge it's at a perfect depth to stand on, squatting slightly, and let the river hold you up. It feels good to be in the embrace of the river. It's a gentle river, here, when it's not in flood. 'The river loves me,' Carrie said, of her experience of coming to it in a time of need. 'It sounds really sentimental but that love is pure and uncomplicated.'

'I found myself calling the river "her"', Sarah said. 'Actually my [PhD] supervisor queried it, like "Wait a minute, where are you getting this from?" and I was like "Oh well, it just kind of came up".'

When I returned to the river at Deep Rock in 2023 even the silt underfoot felt different. David Berman, my favourite songwriter, once sang, 'You can't change the feeling, but you can change your feeling about the feeling.' In January 2023 I realised, or decided, that the silt under my feet and between my toes felt like the softest, plushest velvet. As you walk across it, slowly in the weight of the water, bubbles of gas are released from it and sparkle up the backs of your legs. If you hold your hands at the point of their vanishing beneath the water, they're sepia-toned. It's sentimental, and why not? Why shouldn't we be sentimental about a river? If you think of a river as a sewer, you'll treat it as a sewer, then it will become a sewer. Perhaps if you care for a river instead, the river will in turn care for you.

SWAMP WALLABY

It's impossible not to love a swamp wallaby. I mean obviously it's extremely possible not to love them, especially I suppose if you're trying to grow a garden and they keep grazing on your shrubs (unlike other wallabies and kangaroos, which eat grass, swamp wallabies like to go for leaves that are a little higher off the ground: sometimes you can even see them stretch up and pull down a particularly tasty-looking branch to have a nibble) – but even if that's the case, it's impossible not to love them. This is my book and that's the rules: you've gotta love swampies.

There's a bit of a push lately to rename them 'black wallabies', which is kind of silly because if the name 'swamp wallaby' is a misnomer (they live in all sorts of places, not just swamps, and in fact I'm not sure if they even live in swamps at all), 'black wallaby' is hardly any better: they're all sorts of lovely and subtle colours – very dark brown, grizzled grey and white, golden orange – but black isn't one of them. And a bit like calling *Limnodynastes dumerilii* an eastern banjo frog when you could call it a pobblebonk, opting to call *Wallabia bicolor* a black wallaby just seems like the duller choice. In any case, supposedly the name 'swamp wallaby' doesn't come from these animals' habitat preference but rather from their smell, so it's probably the more accurate name anyway. (I've never been close enough to one to have a sniff, more's the pity.)

You might reasonably assume that there are a lot of wallaby species in the genus *Wallabia*, but in fact there's only one: our friend the swampie. It's out there all on its own. There's something about the teeth, if I recall correctly, that sets it apart. If you live down the east coast of the Australian continent and you've seen any kind of wallaby, you've probably seen a swampie: they're a little shy, but they also like to get out and about during daylight hours, which all things considered is very obliging of them. I've often been out bushwalking and turned around to see a swamp wallaby standing 20 or so metres away, upright and staring at me. If I move they'll bound away; sometimes if I look at them too directly they'll bound away – but if you give yourself soft eyes and you stand still or move gently you can often lock eyes with a swampie for a good few heartbeats. Happy times.

I've seen a few swamp wallabies along the Yarra. As with

most wild animals it's likely that a good few more have seen me. Once you start getting into the outer suburbs of Melbourne and the city fringes of course you get into the kind of country where you might reasonably expect to see a wallaby, but you can find them closer into the city than that too: once I was walking through Banyule Flats Reserve on the banks of the Yarra in Viewbank, only half an hour by public transport from the CBD, and a swamp wallaby suddenly broke cover and hopped across the track in front of me. I can't remember exactly what time of day it was but I'm not a crack-of-dawn kind of a person so more likely than not it was in the middle of the afternoon.

Apparently there are swamp wallabies that live in Yarra Bend Park, in the inner city. I've never seen them, but I've talked to people who have. Someone I know found swamp wallaby poo on Wurundjeri Spur, where I go walking all the time. (You might have a different opinion about this, but I feel like if you've got to the point in your life when people are sending you photos of unusual animal poo, you've made it.) Someone I talked to even told me they'd seen eight of them at once bouncing along the cliff ledges at Deep Rock, which is extraordinary: unless it's a mum and bub, swampies are usually solitary. But Yarra Bend Park can surprise you. Once I saw an eastern grey kangaroo hopping around next to Yarra Bend Road where it passes over the Eastern Freeway. You can see the towers of the city centre from there. In 2021 a deer suddenly appeared in Clifton Hill and made it all the way to Fitzroy, and if there's one deer it's likely that there are more, but I've never seen them. There's a lot of cover along the urban Yarra and its surrounding bush for even a reasonably large animal to hide in.

It's not a wallaby, but case in point: not too much longer before the deer, in 2020, a koala suddenly turned up in Kew, next to the Burke Road Billabong. A friend alerted me to it and I just happened to be nearby, at another of the billabongs that proliferate through the middle Yarra. I hopped on my bike, rode down the Main Yarra Trail and there it was: a koala, leaping about in the treetops. (I know that sounds like hyperbole but I really did see it leap from stem to stem in the top of a small eucalypt and I even have a video of it. It didn't do very much after that.) I have no idea where it came from, no idea where it went to. It just manifested – or at least you could be forgiven for thinking so.

I'm very hopeful of one day manifesting one of the Yarra Bend Park swamp wallabies. To be honest, like most wildlife watching, it's probably just a matter of parking your bum on a comfy seat and having a look for a few hours. The biggest obstacle to most wildlife watching, I've found, is boredom. Which is really just another way of saying short attention span. One time I was in Far North Queensland (yes I know that's a very, very long way away from the Yarra, but let me brag) and I was sitting at a picnic table in a clearing in the rainforest, eating my lunch, and I looked up from my sandwich and there was a cassowary standing not 5 metres away from me on the other side of the table! It's conceivable that if my sandwich had been just a little bit tastier I might have been so engrossed in it that I wouldn't have noticed the cassowary at all.

It's always a delight when you turn around and find yourself face to face with a swamp wallaby. They've got a very sweet face, a little buck-toothed, which makes you want to pat them on the head in a condescending way (though of course if you tried to do that you'd never see that wallaby again). Once at dusk on Wurundjeri Spur I turned around – having just been marvelling at a flock of gang-gangs, not common along that part of the river – and there was a wallaby! Was it? Surely not! But wasn't it? Dark, hunched over, furry ... No. Turned out it was just someone's dog doing a poo. Its owner was just behind it.

Still, you've got a better chance of accidentally seeing a swamp wallaby than most other types of wallaby because of that whole diurnal thing I mentioned. Which is the opposite of nocturnal, which is what most wallabies are – nocturnal, that is. We're diurnal, too (most of us), so that's handy. It seems like a funny quirk of fate that the wallaby with the darkest fur is also the one that's out and about in the sunshine – I don't even like wearing a black t-shirt on a sunny day. I guess it doesn't make them too uncomfortable though or evolution would have sorted that out, as it's wont to do.

And now, to my embarrassment if not shame, I find that I've run out of things to say about swamp wallabies. You'd think I'd be able to natter on for thousands of words about them, but I'm struggling. I think the bottom line is, I just love them – and that's it. I just love them. Maybe you can tell that from the fact that they're the only animal in this whole book (don't fact-check this) that I've called by a nickname. Swampies, I love youse all!

A MIGHTY BEAST OF THE ECOLOGY

When I met Sam Wallman the first thing he said was 'I have something for you.'

'Oh!' I replied excitedly as I saw the poster tube under his arm. 'Is it your "Make the Yarra swimmable" cartoon?'

'No,' he laughed, half apologetic and half delighted as he handed me a rolled-up poster-sized print of his cartoon 'A guide to the Yarra River's beats'. The cartoon winds sinuously through a rainbow from warm orange in one corner to lush violet in another – a 'melty rainbow map' in Sam's description – tracing a history of illicit sexual encounters along the river's banks: 'Lie back and imagine an ancient Gondwanan forest', one panel advises, the text encircling two men engaged in oral sex as the river bends past them. Later in our walk along the river Sam would recount his experiences as a gay man, cruising 'many times with deep intimacy with strangers by this river, picnics and drinking, then just trying to understand the physical attributes of it, spiritual attributes, the history of it, the industrial history as well – there's so many aspects that I love about it'.

Sam was modest and self-effacing about his work during our forty-five-minute conversation, but his intricately detailed cartoons display deep thought and depth of research, as well as a welcome political and social earnestness: time and again after reading one of Sam's cartoons I find myself filled with hope at the possibilities of the future. 'If we were able to swim in the river,' his swimmable Yarra cartoon says, 'that would represent the cherry on top of the restoration efforts. Because it can't be healed in isolation. The river is a drill-core sample of the broader ecological state. It's a barometer, a feedback loop and a vein – supporting the lives of millions of creatures and countless ecosystems.'

At the top of the cartoon a person – short, dark hair; small gold-hoop earring – swims freestyle in the river. The water flows around and into them, the figure's open mouth like a bay drawing in the river. I told Sam how much I liked that detail, of the river flowing into the swimmer's mouth. He laughed and said, 'I know, my friend was like, "Man, don't have it going into the mouth, that's too visceral." I was like, "If we can't even cop the idea of it going into someone's mouth …"' He trailed off, the question of the limits of what we're willing to imagine left hanging in the air.

Sam's passionate about the river: at one point in our conversation he recounted how 'on the weekend this fella was talking about how dirty the river was and I jumped to action straight away'. Adopting a comically distraught voice, he relived his castigation of the man: '"It's silt! It's floating silt! Leave it alone!"' I told him about how I'd had to unlearn negative attitudes towards the river since moving to Melbourne. 'Oh, it's so offensive!' he replied. 'Even from a settler perspective, this is a mighty beast of the ecology, an ancient part of this colonised land. It's kind of insult to injury to shit on it, really.'

He asked me about my upbringing, so I told him about growing up in Canberra, and swimming in rivers there because the sea was so far away. 'Yeah I was a river kid,' he said. 'I'm from Geelong and the Barwon River was my ... as a little kid I'd ride my bike by it every day, then as a teenager I'd smoke weed by it every day, it had different appeals to me through my life. I remember as a teenager my mum and her friends being like "Do you prefer the ocean or the river?" – they were chatting among themselves. Well both are amazing, it's obviously not a binary, but I do feel myself drawn to rivers.'

I'd asked Sam to meet with no real idea of what we might do or where we might go, other than it seemed apt to meet somewhere along the river. We ended up crossing the Johnston Street Bridge from the Abbotsford Convent and then stepping onto the walking track that follows the left bank of the river – part of a longer track that takes three or four hours to walk in its totality and is my favourite urban bushwalk in Melbourne or just about anywhere else; the same track I'd walked back and forth so many times searching for Salvatore the seal. Sam had never been this way before and he was immediately thrilled – he jokingly described the path as 'the DVD special feature' of the river. Almost immediately we spotted a pair of tawny frogmouths roosting below the track; a little further on we passed by a man taking photos of a kookaburra. We talked about how cities can be biodiversity hotspots, often unbeknown to their human inhabitants. 'You can take heart from that,' Sam said. 'It's all a kind of resistance, really. All these animals.'

We spoke about the restoration of the river, the revegetation that's occurred in the last few decades. 'It's pretty remarkable,' he said. 'It's so recent, our parents would remember it being deeply maligned and also totally locked up because industry would have been right up to the banks. I feel like we even take for granted the fact that we can even just walk along it without a cyclone fence in our way ... Like it's a little glimpse of the commons, really … I feel like in the same way that people understand something about private property being a really foul concept in their understanding of the beach, even my dad gets so arced up if council puts some awnings on the beach or something like that, there's a bit of that with the river too, it allows people to imagine a different sort of public ownership.'

I asked him about his own relationship with the river. 'If I feel mentally unwell or anything I find movement has always helped,' he said, 'so cycling along the river has always been my default if I feel crusty or crazy or anything … you kind of lose your body in some way, even lose your mind if you're sweating and your heart beats fast enough, it's great.' I asked him if he'd ever swum in the river. 'Yes, yes,' he said. 'Intentionally and unintentionally. I swum in it actually not far from Galatea Point years ago and didn't get sick thankfully. I've swum in it many times at Laughing Waters. That's cheating, though, 'cos that's very far upstream. Warburton – icy swims up there. My brother and I were up there last year in between Melbourne lockdowns, and we were swimming in Warburton and we found an old [tobacco] pipe in the margins, on the side of the banks, so very cool. Now we share this as a memento. And then I was a member of the kayak club, the INCC [Ivanhoe Northcote Canoe Club], and obviously stacked it a couple of times learning. Which is always what I remind my friends, 'cos I would always take them out with me kayaking and be like "You will fall in at the start", and if you tell people you want to go swimming in the Yarra they'd never do it but they're somehow willing to fall in accidentally, so see? You didn't get crook. Just keep your mouth shut ... It's an act of love to swim in it I reckon at this stage, rather than a natural act, but I think that will change.'

It can also be an act of pleasure, and of leisure: there are no swimming lanes in a river, no carefully measured 50-metre lengths, no lines marked on the bottom. In the UK, swimming in a lake or a river or the sea is sometimes called 'wild swimming', but a more inviting name might be simply 'recreational swimming'. In the last couple of years of working from home since the start of the COVID-19 pandemic I've found that I've been able to reclaim some of my leisure time – the vaunted 'eight hours' recreation' that increasingly seems to have become the forgotten part of the eight-hour day. Of course for many people in precarious work eight hours of recreation – let alone eight hours' rest – remains an impossible aspiration, but I want to believe that the glimpse of another way of living that we got at the start of the pandemic still remains, and can still be seized upon if enough people realise it.

'Another part of the eight-hour movement that I try to

remind myself about,' Sam told me, 'is that they wanted time to organise as well.' As he said this we stopped to try to help an ant that was curled up on the dusty track and in danger of being stepped on. 'I'm always thankful that they've sent us this eight-hour day through time, more leisure time, and then I remember that they also wanted us to organise in that time as well. We also have the time to fight for the river because of the eight-hour day too.'

Thinking about recreation along the Yarra, I told Sam about the community of swimmers that has grown up around Deep Rock, inspired by the Yarra Yabbies; how you can go there in the morning before work or the evening after work, or on the weekend, and there will be people having a dip. 'And a chat as well,' Sam pointed out, 'because that's another lever. You can fight people's loneliness in this [swimmable Yarra] campaign too, and if someone did get sick or if there was visible pollution there'd be a natural base to push back and call their MP or whatever ... Imagine if we had communities like that dotted all along the [river], like if each neighbourhood had a little swimming club.'

At the start of our conversation Sam borrowed the *Simpsons* neologism 'crisitunity' – something that's a crisis, but also an opportunity – and at the end of our conversation he revisited it: '[The river's a] Rorschach test really, you and I probably look at it differently, let alone some Tory ... And look, we've been saying if they want to make it cleaner for their property prices that's okay too, whatever gets us there. We're only gonna fix it if we deal with the structural shit, though, obviously, because it's so connected to everything. That's why in the poster I drew all those forces flowing into it as streams 'cos that is how it is, all these different issues are in that river; it's not polluted by the industry but it's polluted by its other stuff ... I'm so tempted by the shortcuts. But no, if we do just make it swimmable we won't have used this crisitunity.'

I often think of the English poet John Clare, who documented the privatisation of common land in the earliest days of industrial capitalism. The river, too, like other urban rivers, has historically been privatised: even if the water itself was publicly owned, pollution from private industry so befouled it that it became effectively fenced off, unwelcoming and unloved by Melburnians. But the Yarra is not that river any more – especially in the wake of its popularity during Melbourne's lockdowns. 'Yeah, it's the commons,' Sam agreed. 'It is a portal into some other way of having a city and sharing a space. 'Cos people get along [by it] ... So maybe this is the frontier of the fight against capitalism still. [This is] such a cartoonist's thought but I'm like, "How do we link arms with the river and picket line or something?"'

Sam's own relationship with the river continues to change and grow: 'It pierces the veil somehow,' he said as we walked along the track above the water. 'Even a month ago my partner and I were dog-sitting in Heidelberg and we went to some brewery and we were coming back after dark with the dog and we decided to walk along the river, and I guess I'm my mother's son, I usually don't go to the river at night. You hear one story of someone getting garrotted while cycling ...! I don't even know if that ever happened, but "stranger danger", I'm just probably not going to go down ... But we walked along and it was just so beautiful in the darkness, just the low light and the different things that the low light highlighted, and the sound – it's very loud ... I saw it from a completely different perspective and it just unlocked it in a new way ... We don't live at a time when people have a very sacred outlook, but I feel like there's some sublime character to the river.'

We finished our walk, accidentally but appropriately, on Yarra Street, completing a loop from the convent and back again. We said a slightly awkward farewell, and in parting Sam concluded, 'Everyone loves [the river] and respects it, they just don't know it yet.'

SNAKE

The biggest and most beautiful tiger snake I've ever seen was sliding out of the Plenty River onto broad, sunny rocks where that tributary joined the Yarra. I was on a footbridge several metres above, which gave me a great vantage point that was mutually advantageous: the snake and I were too far away to get in each other's way.

Every year when the weather warms up the dog-walking community in my suburb starts getting excited about snakes: warnings are sent out of snakes seen and narrowly avoided, of dogs nearly being bitten. I never see them and I'm probably the only dog walker in the neighbourhood who every year desperately *wants* to see snakes.

Snakes are great. I'll tolerate no snake badmouthing here. In my stomping about the bush I've posed more of a threat to snakes than they have to me, having nearly trodden on two tiger snakes on the banks of the Yarra. One of them was the only tiger snake to have ever threatened me in any kind of way, flattening its neck to look bigger – and even then it was hastily sliding away from me as it did so. I'd been enjoying a lovely walk in the sun along a rocky path just downstream from Fairfield Boathouse and, not paying attention to where I was treading, had put my foot down 5 centimetres from the snake. I was only alerted to its presence by the sound of it moving away through a chicken-wire fence. Under the circumstances I think it was remarkably tolerant of my blundering.

People think that tiger snakes are fierce and scary just because they're extremely venomous, which I suppose is fair enough as far as that goes, but if I were to accidentally tread on a tiger snake it would probably think of humans as being extremely stompy and aggressive. Being venomous and capable of killing a human does not in itself make a tiger snake an unpleasant snake. Every single tiger snake I've ever encountered has only been interested in doing one thing, namely getting the hell away from me.

And having lived in Melbourne for nearly twenty years, I've encountered more tiger snakes than I have any other snake species – even though it hasn't been nearly enough of them for my liking.

They like to eat frogs, you see, which means that they like

to hang out where frogs hang out, which for the most part is in or near water, which means creeks and rivers and wetlands – and Melbourne has a *lot* of those. Lots of water, lots of snakes that like to live near water (or sometimes *in* water: once, many years ago, I was hiking with my dad in Tasmania and we were crossing a creek where a tiger snake was hanging out, and it escaped by submerging itself and sliding under a rock on the creek's bottom; we watched for what we both remember being about ten minutes and the snake didn't re-emerge).

Speaking of frogs, on any given night throughout the year you can hear common eastern froglets calling *crick-crick-crick-crick* from any pond or billabong along the course of the Yarra and all its tributaries. They're minuscule – which I guess is why they only qualify as 'froglets' – but they make up for it with their voices. Depending on the time of year you'll also stand a good chance of hearing southern brown tree frogs (yes, there are tree frogs in southern Australia!) and pobblebonks. Very dull people sometimes call the latter species eastern banjo frogs, but try out both names. Say them out loud. It's pretty obvious which one is the more fun to say.

Fair enough, these frogs *do* sound like somebody plucking a banjo string. But they also sound like 'pobblebonk', so accuracy doesn't really come into it as a deciding factor. What

I find interesting, though, and what I'd write down as my first question if I were ever to take a turn and start hosting pub quizzes, is that the word 'pobblebonk' actually describes the sound of *three* pobblebonks: the first one says 'po-', the second one says '-bble-', and the third one says 'bonk'. All in quick succession: *pobblebonk*. Sometimes I like to spend my time wondering (by which I mean, waste other people's time by asking them) if there are any other animal species named after the sound made by more than one individual of that species. I can't think of any off the top of my head, but if you can I'd love to hear about it. (And I can't speak for anyone else, but I'd also be perfectly happy with a frog that was just called a bonk.)

I don't know if tiger snakes have a preference for one type of frog over another, but pobblebonks would make a more substantial meal than common eastern froglets. Maybe that's why you hear more common eastern froglets than you do pobblebonks – not that the latter are rare. There are other frog species along the Yarra, too, and every year I vow that I'm going to learn their calls so that I can identify them by sound, but every year I find that my brain has only committed to memory the three species: common eastern froglet, southern brown tree frog and pobblebonk. Still, it's a delight to be out along the Yarra on a mild evening and hear any frog at all.

It's also a delight to see their predators the tiger snakes. Every time I see a snake I stop and admire it; I can't help myself. Partly of course I'm also stopping and watching it, because despite my enthusiasm I am acutely aware that a tiger snake could kill me if it was of a mind to, and if there's a snake around it's better for you and the snake both if you're aware of its presence until either you or the snake has moved on. Ideally I think it should be the human who moves on, but I guess it depends on the circumstances. I probably wouldn't move out of home if a snake decided to take up residence – I reckon I'd feel like it was fair to expect the snake to move on instead. I guess you can take that both ways, though, when we end up in the snake's back yard. Anyway, as you've probably guessed by now I won't hear a bad word against snakes. (Or frogs.) (Or most animals, I guess – let's be honest.)

I used to be really into birdwatching and to be honest I still am but a few years ago I realised that if I wasn't careful I'd be branded a 'bird guy' so I bought a bunch of field guides and started getting passionate about other animals, too. I mean that's not really how it happened, but I did get field guides to reptiles, mammals and butterflies all within the space of a few months. I also own at least three bird field guides so I guess I'm mostly into birds still. But I do love all the animals. Anyway, where I'm going with this is that when I got the reptiles field guide I learned for the first time that there are two species of blue-tongue lizard in Melbourne, and then when I went and looked at photos I'd taken of blue-tongue lizards (just on my phone – loving all the animals is one thing but becoming a wildlife photographer is a whole other threshold I will not cross) I'd actually seen both species, I just hadn't known it at the time.

I didn't have that realisation for a few months, though, so for a while I was obsessed with seeing the 'other' species. I still am, to be honest. Not that I don't get excited about seeing 'normal' blue-tongues, and I still look forward to seeing my first one of the season every year when the weather starts to warm up. But I've seen lots of them (knowingly) so they're not quite as exciting.

I've sort of painted myself into a corner here by not telling you what the two different species are, which means that the last paragraph was unnecessarily cagey in a way that it wouldn't have been if I'd just told you from the outset what they are, so I'll tell you now. Better late than never (much like when I found out about that second species). Here you go:

The species that I knew about already, and that you might be familiar with, what you might think of as the default blue-tongue if you live in south-eastern Australia (unless you live in Tasmania, in which case the other species is your only species), 'blue-tongue classic', is the common blue-tongue. I'll admit it's a bit of an anticlimactic name after all that build-up I gave it. You'll recognise it (again, unless perhaps you live in Tasmania): it's a big skink (blue-tongues are skinks), chunky, with short legs, the deep violet-blue tongue of course and, most distinguishingly, broad horizontal bands across its body. We'd call them hoops if they were on a football jersey.

The other species (unless you're in Tasmania, in which case it's the only species) is the blotched blue-tongue, and the name tells you how it differs from the common blue-tongue: instead of hoops it has big pale spots all along its body, either side of a dark line running along its spine. (I'm not aware of any football jerseys that we could compare this pattern to so you might just have to google it to get the idea.)

Both species are found along the Yarra. And in plenty of other places, too, but the Yarra is what's pertinent here. Common blue-tongues in particular are often mistaken for tiger snakes because of those hoops (which we'd call stripes if they were on a tiger snake – I should possibly abandon this whole 'hoops' concept). As with a lot of other reptiles you'll often only know that blue-tongues are around because you hear them moving away from you through the undergrowth. It's very frustrating, to be honest. If you hear it you can usually tell a lizard from a snake even if you don't see it by the fact that a lizard will run a few steps and then stop, whereas a snake will slide away. Sliding makes a continuous, steady noise, while running, requiring the placing of feet on ground, sounds a little more disjointed – though when it comes to blue-tongues the placing of feet on ground cannot always be relied upon: more than once I've seen one making a hasty getaway with its back legs dragging limply along behind it, just there for the ride.

By the way, it's a bit of a misconception that reptiles are only active during the warmer months. People are sometimes surprised to find tiger snakes sunning themselves on particularly, well, sunny days in midyear, June or July. I often feel like it must be tough being a reptile in southern Victoria: it's not exactly warm, and the sun is usually more absent than present. But I guess they get by. Some reptiles – like the blotched blue-tongue – are even specially adapted to cooler climates. Which presumably is why the blotched blue-tongue is found in Tasmania but the common blue-tongue isn't.

The lower Yarra is also home to a small population of water dragons, a species native to eastern Australia – but not to Melbourne. It's thought that the population here is derived from dumped or escaped pets: the nearest natural population is in East Gippsland. If you go walking through parts of Studley Park you can sometimes see them. Sometimes you can see turtles, too: common long-necked turtles and Murray River turtles. These are both native to Victoria as well as elsewhere in Australia, but not to Melbourne – like the water dragons, it's likely that their presence in the Yarra and its associated bodies of water is as a result of the pet trade. (None of this is my own knowledge, by the way: I learned it from the field guide I bought, so don't give me too much credit.)

And I haven't even started talking about all the other snakes besides tiger snakes! One afternoon in Plenty Gorge Park, near the Plenty River, which is a major tributary of the Yarra, I spent a lovely fifteen minutes or so eye to eye with a delightful eastern brown snake. I live in hope of seeing a copperhead one day. (I made a bit of a fool of myself by declaring with undue confidence to all of Twitter that the aforementioned eastern brown was a copperhead, based on a single identifying characteristic mentioned as definitive in my field guide that turned out to be, on the snake I saw, just a shadow. No matter, lesson learned.) There's a dazzling variety of skinks that live along the river, too. The point is, and you've probably picked this up if you've read this far, there's a lot out there and you might not get to see it all but it's a lot of fun trying.

WHAT IT ACTUALLY TAKES TO CARE FOREVER

Anna Ridgway had been reading *All This, and Heaven Too*, Rachel Field's 1938 novel – perhaps better known for the film adaptation, starring Bette Davis – which tells the story of Field's great-aunt Henriette, a governess who was implicated in the murder of her employer's wife. When we met – not for the first time – in early January 2023, Anna recounted a scene from the novel in which Field reflects that Henriette, having 'grown used to down pillows and fine linen and upholstered furniture ... had not realised how insidiously such comforts could become necessities'.

The question of what are comforts and what are necessities was something that Anna had been pondering. 'It's been interesting to navigate the defensiveness of people who use watercrafts and are jolly well going to use them because "I have a need to use it and I need to get out of the confines of living in the city and get out on the water"', she told me as we walked along the Main Yarra Trail between the Abbotsford Convent and Dights Falls. 'This recreational activity has come to be seen as something one does ... The right to do it becomes muddled with the responsibility to take care and self-restraint in vulnerable areas.'

Anna was at pains to point out that she's not against people using kayaks in the Yarra outright, '[But] when you politely put to them the fact that [according to Kristin Otto's book *Yarra: A Diverting History*] the Yarra River has no natural beaches left, they were all destroyed, and that watercraft landing and beaching zones need to be done very, very carefully because if they're put onto the ground [in highly vulnerable, erosion-prone areas like the lower Yarra below Dights Falls] then they cause catastrophic erosion which is almost impossible to reverse ... you get this very interesting defensiveness.' She invoked St Augustine's famous words, 'Lord, make me haste but not yet', and gave them her own spin – what she called the modern equivalent: 'Make me environmentally responsible, as long as it doesn't inconvenience my lifestyle.'

Anna is not someone who tends to hold back in her opinions. The small organisation she's part of, the Abbotsford Riverbankers, takes a very careful, long-term approach – a nurturing approach – to restoring a severely degraded section of the lower Yarra.

They learn as they go and apply those learnings both to their own practices and, through direct public engagement with the many people who pass through their area of operation along the Main Yarra Trail and through their forthright and passionate Instagram account, to the wider public and anyone who cares to listen. On that January morning Anna recounted another part of Field's novel: '[Henriette is] an artist so she's very intensely aware of everything around her and life pulsating and all the minutiae – and the phrase [used to describe her husband] is "But Henry just saw things as vague mounds of green", and that certainly pertains to how I would say most people look at this, in this case the Yarra.' Anna told me about a Master's study she'd been part of by Ann Seward of RMIT into 'plant blindness' – that is, 'an inability to notice plants in nature, appreciate them or value their importance to human affairs, and a tendency to rank plants as inferior to animals.'

'You can't change behaviours until you get the subtle shift in mindset,' she told me. 'And you may be an overall vague kind of person, but at least within that and within your own wiring you can start to look at [the river] differently and then gradually work yourself towards the different behaviour changes that we see.'

I often go to French Island, in Western Port, east of Melbourne. If you look at it on the map you might think there's not much there, but if you talk to any of the hundred-odd locals you soon learn that they know every inch of the island inside out, and every tiny part of it has a name that will never appear on a formal map. I've walked along the Yarra with Anna twice, and each time I've been reminded of that. The part of the lower Yarra floodplain the Abbotsford Riverbankers have been working for years to help heal is only a few hundred metres from end to end, and walking the length of it will take you only a few minutes if you're just going from A to B – but walking it in the company of Anna is a slow, deliberative affair, full of stopping and observing and considering. It's the Riverbankers way. 'We put a lot of emphasis not on the size, but on what it represents,' Anna told me. 'A non-park, urban thoroughfare – what can be done, and how that can be replicated again and again. Yet in our cultural mindset, behaviours, practices and the narratives that go with them, the focus is overwhelmingly numerical – kilometres square. How big our area is ... toting up facts. Counting numbers. Competing against others for how many numbers of things you can find. Also we're in the heart of community. We get to talk, thrash out ideas, explain, challenge and be challenged right in the midst of our floodplain community. People who focus on geographical size tend to disregard this.'

We paused at a particularly steep area of riverbank. 'It sort of represents just how psychologically difficult it is to do this,' Anna said of the area, 'and how you have to hold your nerve in the context of this unbelievably catastrophic invasion of, particularly, Weeds of National Significance [a list of plants considered by the federal government to be particularly invasive and harmful to the environment] like the Madeira vine up there,' she pointed, 'and of course just how fragile these riverbanks are structurally ... under those vague mounds of green.'

Anna gestured towards the slope of the upper riverbank we were standing next to, which was covered in invasive vegetation, the trees growing on it smothered in Madeira vine. '[A]ll that ecology that was ripped out 170 years ago, it hasn't really recovered, so when the floods [in October and November 2022] came up literally to where the trolley is there ... it did a lot of gouging out, which reinforces why we've had to keep all of these grass weeds around here because basically they're holding the bank together. But it makes what we do still extraordinarily difficult because we keep having to hold that [vulnerability to flood-based erosion] at bay while holding our nerve with opening this out to rehabilitation but knowing that when floods come – and they'll come more aggressively and more often in the future – they could continue to gouge out. And these Madeira vine weeds are still very, very much in the soil so they'll keep coming out and coming out and coming out ...' The long-term struggle against Madeira vine, Anna explained, can be psychologically overwhelming to people who are committed long-term to ecological healing. A mindset of forever care is required of anyone who wants to join the Riverbankers. 'This is not for the "stick in a plant and pull out a weed" ad hoc volunteer' Anna said.

I'd first contacted Anna nearly a year earlier. I'd intended to arrange an interview but instead she invited me to join her and a young volunteer on a site walk: an induction for new Riverbankers. She took us through the various parts of their area of operation, and then instructed us through planting some tubestock sourced from the nearby Victorian Indigenous Nurseries Co-operative (VINC). She explained the Bradley method of revegetation that the Riverbankers have adapted to their context: '[It's] about helping nature heal itself and doing that by starting with what's already there, what's already existing. You clear and you secure around that and you're weeding or you're planting or whatever, but you don't go into weed zones – you leave them alone because they're just going to grow back and back and back and back, there's no point. So start with what there is and take care of that and allow that to gradually spread further and further.' She guided us through planting young sedges in the flood zone, hollowing out a cup in the earth to hold them so that when the floodwater came it would flow in and flow out again without tearing at their roots. She explained about the various weeds that riddled the area, of which Madeira vine was – and is – the most virulent. Digging just below the surface of the soil on hands and knees in one small area we found dozens of tubers – and Anna warned us that any attempt to simply rip out the Madeira vine growing all around us would dislodge thousands of aerial tubers from each of which a new plant would grow. 'It would also fail to remove the underground tuber colonies the vines grow from' Anna said, 'while ensuring that disturbing the soil would stimulate even more growth from these rainforest plants.'

Madeira vine can so heavily infest an area that the weight of it is enough to bring down a tree. 'So we've decided strategically to leave the weeds alone and just brush-cut them,' Anna told me on our follow-up walk in January 2023, 'because they're holding the bank together, but when you brush-cut because it's still disturbing the soil you're encouraging these guys to grow even more. So again it's psychologically incredibly difficult because you have to be prepared that the rate of growth is going to double while you're doing what you need to do. When it comes to loving and caring for the urban part of the Yarra River, this is as difficult as it gets, and when it comes to mindset change and behaviour change – yep, this is really getting to the kernel of what it actually takes to care forever for these riverlands.'

Anna doesn't back down from drawing attention to the way

that people sometimes crash through the environment without considering their impact – though 'We all do this to some extent' she says, 'and I'm 100 per cent realistic that this is just human nature … I don't [draw attention to] it by "shaming" people. I simply point out the actions and the mindsets, and make sure I put myself out there to build bridges with people very different from myself.' To see the delicate recovering ecosystem of a severely degraded part of the riverbank – one with no remaining remnant native vegetation – trampled underfoot causes her deep and profound pain, but so do 'guerilla' individuals who swoop in and pull up every weed in sight without considering the context or the consequences. The Riverbankers have sometimes come to their work area to find that it's been sprayed. 'In recent years,' Anna told me, 'this was done by locals trying to get rid of weeds, but they did it behind the Riverbankers' back. The consequences were bank slump and collapse.' Anna was doing site inspections on two occasions in December 2020 and January 2021 and her video footage of the bank collapsing above and below the Trail below Turner St, submitted to Melbourne Water, Parks Victoria and City of Yarra, led to reconstruction of a section of riverbank and repairs to the damaged drainage system above. 'Without having any literal evidence,' she told me, 'I would say [part of it is] malicious spraying to weaken rehabilitation areas … These kinds of anonymous, perverse, malicious behaviours have been an ongoing issue for us … I know that some locals don't like the thick weed grasses growing at the [Main Yarra] Trail edge' Anna told me. 'They're brush-cut regularly, but some people spray it. But they've also sprayed along rehabilitation areas, like [in 2022] just before the historic October–November floods, freshly weakening the area.'

Sometimes the public attitudes Anna and the Riverbankers encounter are more prosaic. The area of the Main Yarra Trail they work along is particularly low, and so they take the force of seasonal flash floodwaters that come surging from the convergence of the Yarra and the Merri Creek and down the artificially straightened, shallow Yarra channel below Dights Falls. At either end are boom gates that swing shut to close off the trail – or at least that's the intention. '[These gates] got closed during floods even when the water was down, and of course, speaking of luxuries being necessities, [it was] "I'm bloody well going to go through because I want to ride my bike through here and I can't see water on the trail so I'm going to go through here even if the gate says don't go through" –' Anna indicated a worn-down patch next to the gate, once vegetated but now bare earth in a semicircle of foot and bike traffic '– and so of course this just keeps happening multiple times a year during multiple flood seasons, so they will walk through here or cycle through here or bring their little kiddy cart in here in this most vulnerable of areas and we just have to watch them go through.' She sounded tired. 'And so for me it's a case of managing panic attacks and managing the anger and the frustration and just seeing it over and over and over and over again and you can't do anything except to go in and put mulch down and put rocks down and just try and protect it.'

If you spend enough time in Anna's company it becomes clear that her directness of speech is a manifestation of how deeply she cares. 'It's an ecosystem of managing your emotions,' she told me, 'and supporting your team who are at the pointy end of all this to manage theirs, and to do so in public, and to communicate judiciously and then to weave that into your forever message of floodplains care and what it means and what it takes and the little powerful things that you can do. Like during flood season, do what the Wurundjeri-Woiwurrung people would do and go to higher ground; there's a bunch of stuff you can do up there, there's a lot of beauty, there are safer areas, leave it alone [down here] to dry and heal.'

I ventured the thought that perhaps people were also learning to engage with the river more than they had in the recent past, and that perhaps recreation could help with that. The Abbotsford Riverbankers operate only a ten-minute walk from Deep Rock. Anna wasn't entirely convinced. She agreed that recreation helps people engage with the river, but she wondered about the quality of that engagement: what do people understand about the river with which they're engaging, and its surrounding environment? She pointed out that different parts of the river needed different approaches from people using it: a consideration of the specific local environment before just jumping in. 'Deep Rock itself can be a good place to swim,' she said, 'because it's wide, there's infrastructure ... but it shouldn't be used as the template for every single place, just as boat launches with infrastructure can't be the template for highly erosion prone areas lacking infrastructure or other controls.'

I told her that it had occurred to me that a lot of the public education she engages in – whether stopping passers-by to explain to them why some areas of the riverbank are off limits, or taking to Instagram to point out people who trample over delicate young plants – is about trying to get people to think in a long-term way rather than just about what they could see here and now. 'I'm very glad that you've perceived that,' she said, 'because I've been working very hard on that being the message, and even within our own team it's very, very difficult to change that mindset. So when people – as they do – roll on out of the team onto the rest of their life, and there's a sort of "Oh I'm so sorry I can't be here and reveg" … [I think at least] you understand the long-term thing so hopefully you'll take that on your journey with you.'

It's not just public attitudes that the Riverbankers have found themselves trying to guide. Their first students are themselves: 'I'm not an ecologist,' Anna explained, 'I'm a journalist and a teacher – but you have to gather the bits and threads of the knowledge and experience and learning and advice when it comes to planting down in these [flood] zones.'

Their partners, too, have required some guidance: 'Now we get our stuff from VINC nursery, and they've been fantastically helpful but it's taken a while to build a dialogue around what our context is. So it started as "Okay, we need this plant, it's the ephemeral zone or the emergent zone of the river, it's flash floods and … all that kind of stuff" and the answer from our interlocutor at VINC would be "Plant this thing, they love getting their legs wet". I think it might have been *Geranium inundatum*. "Okay, fine, we'll plant *Geranium inundatum*." We ordered a bunch and planted them and of course they were completely destroyed by floods! So we had to come back and say, "Look, this is what happens, this is the context, we need a different kind of advice from you guys." And that took a while … So it's kind of benefited them as well as us because … they'll then be able to pass that on to other possible organisations that are dealing with these riparian areas. But [it's] very, very slow burn.'

We stopped to look at a recovering acacia growing out of the riverbank, between the trail and the river. After the recent floods it was strung with litter, plastic tinsel on a struggling Christmas tree. 'People are conditioned to look at litter in trees and think, "Oh my god, litter in trees," automatically that's 100 per cent bad,' Anna observed, 'again without understanding that these trees are effectively litter traps and they are preventing the rubbish from going further downriver.' It was an unexpected thing to hear from somebody so passionate about the riverlands – but instantly I realised that it was of a piece with Anna's holistic way of looking at the environment around her: not the forest or the trees but the forest and the trees.

Some may hear Anna talk about the Riverbankers' work, or read the group's Instagram posts, and see an austere world view, one in which life is struggle and toil. 'That actually is my view,' Anna told me. 'That's the world.' But it's possible, if you're open to the message, to recognise in it something more complex: a selfless world view, which seeks to put the self in service to another – in this case, the Yarra. It's a world view I grew up with, too. It can be hard, sometimes, but also endlessly nourishing and unhesitatingly nurturing – deeply and profoundly generous, in fact, and even hopeful. 'Life is struggle and toil,' Anna said. 'But that's inextricably bound with joy, exuberance, fun. I certainly consider myself both austere and exuberant – life loving, appreciating simple pleasures – and these qualities resound throughout the organisation.'

We looked at the litter-strewn acacia which at the end of its short life might collapse onto the bank, or into the river. 'Snags matter,' Anna told me as. 'They're the trees, branches, root masses, woody debris found in and at land-water interface of rivers. They protect [those interfaces] and provide micro ecologies. Abbotsford, Richmond and Burnley are about the last places in the Yarra you can see them, because they'd be removed from the CBD area for visual amenity and safety ... and they are contributing to holding the show together, but again we're conditioned not to like that, and so part of the conversation is "What about that ugly, dead tree, do you think somebody might come and move that dead tree, when are they going to come and remove the rubbish out of that ugly, dead tree?" without understanding just how important a function it [has].' The river flowed behind the straggly branches of the tree, past a mosaic of fallen branches and logs amid the vegetation. 'This is precisely what the river needs,' Anna said, 'so to be able to stand here and look at this here is really a bit of a minor miracle.'

RAKALI

In one of the lockdowns of 2021 – lockdown six, I suppose, but honestly who can even remember now? – I became obsessed with rakali. It was an obsessive time; how else to get through the days but by fixating on something? I'd just had a book about quolls published – maybe you've read it? No stress if not – and I was thinking that perhaps I ought to have written about Australian native rodents instead because let me tell you, they're amazing, and so I decided that I really, *really* wanted to see a rakali.

When I was a kid they were called water rats. There's an ongoing push to rename species with names from Indigenous languages, and on a few occasions it's stuck – the rakali being one of those occasions, because it was thought that there was too much stigma around the word 'rat'. I'm equally in favour of honouring original names that have been in use for thousands of years and of rehabilitating the bad name of rats everywhere, so it's probably a good thing that I don't decide these matters because a decision probably wouldn't ever be made – but anyway, someone, somewhere (or, more likely, a bunch of someones in a whole bunch of somewheres) did make a decision, so it was out with 'water rat' and in with 'rakali'. Whatever you call them, though, they're amazing. All of Australia's native rodents are.

If you weren't aware that Australia even had native rodents, you are in for a treat. We have dozens of them! And yes, they really are rodents. They're not marsupials any more than the introduced house mouse or black rat are; they're genuine placental mammals like you or me or your dog or your cat (or the introduced house mouse in your kitchen cupboard or the introduced black rat in your garden shed, for that matter). And they're just as native to this continent as any kangaroo or possum or platypus or antechinus, the latter of which are often called marsupial mice but aren't mice at all because they *are* marsupials, and marsupials by definition can't be mice because mice are placental mammals and not marsupials. Funnily enough, there are hopping mice that hop like kangaroos – except they're mice. With placentas and everything. There are even hopping mice in Victoria! Though not along the Yarra.

Native rodents that you *might*, if you're very lucky, see

along some parts of the Yarra include swamp rats and bush rats. They're both lovely. Once I went to Cranbourne Botanic Gardens in Melbourne's outer south-east because there are wild bandicoots there and I saw the bandicoots but I was just as excited when I saw a pair of swamp rats running around in the undergrowth and gnawing on sedges: I watched through my binoculars as they held the leaves between their two front paws and gnawed methodically along the whole long length. Once I thought I saw the body of a bush rat on the bike path along Merri Creek but it turned out it was just a black rat with part of its tail missing. If you ever see a rat in south-eastern Australia, look at its tail first: if it's long, as long as its body or even longer, then it's a black rat. If the tail is shorter than the body then you might just be looking at something exciting.

And if the tail has a white tip, and the rat that the tail is attached to is enormous, I mean properly giant, and also both rat and tail are in or near the water, then you've got a rakali. Rodents Of Unusual Size do exist, friends, and they're gorgeous, and they're in the Yarra.

My lockdown obsession began because a friend of mine saw a rakali near Studley Park Boathouse and posted the photo to her Instagram and Facebook accounts. But I've always loved rakali. When I was going to uni in Canberra I used to detour on my rides home around the foreshore of Lennox Gardens, on the banks of Lake Burley Griffin, because there was a rakali that lived there and more often than not I'd see it perched on a rock or swimming under the water. To my great embarrassment I've lived in Melbourne for nearly twenty years and I've never been to see the penguins in St Kilda but if I ever do go I can guarantee that I'll be just as excited about seeing the rakali that everyone's told me live there as I will be about seeing the penguins. Maybe even a little bit more excited.

It's not that rakali are rare. They used to be: well into the 20th century they were hunted for their fur. I once met a man who told me his father had made a living killing them. But they're found across a wide swathe of Australia, and in New Guinea too, and they're highly adaptable. Give them a bit of land to live on (or in – they can dig burrows), and enough prey animals to eat, and a bit of water, and they're happy. Doesn't even matter if the water's salty or fresh. They like to use their sharp rodent teeth to bite through yabby and mussel shells and pretty much any other animal they can wrestle into submission, and if there's a handy jetty or pier or pontoon to crouch on while they eat it then they'll readily make use of that. Their fur – that once-coveted fur – is thick and luxurious to keep the water out, and beautiful too: dark chocolatey brown on top and gorgeous golden yellow underneath. Their feet are partially webbed. Their nostrils are on the top of their snout, so that they can breathe while they're swimming along the surface of the water. And they can dive, and hunt after things underwater using their long, sensitive whiskers. Up in the Kimberley they've even figured out how to flip cane toads over and eat them without getting poisoned.

And they're huge. I mentioned that. Sixty centimetres or so from nose to tail tip. For reference, that's more than twice the size of those black rats living in your shed. They're often referred to as the otters of Australia, because everything on this continent has to exist with reference to something from the northern hemisphere, and it's not because they're cute as a (very large) button (though they are): it's because they swim up and down waterways eating anything that moves.

They're often nocturnal, but not always; they're also crepuscular, which means you can see them around dawn and dusk. And sometimes you can see them out and about in the middle of the day. Basically they're an animal perfectly designed to tempt slightly obsessive writers into wandering up and down riverbanks for hours like a character from a gothic novel, casting meaningful, longing gazes at the water, wilfully misinterpreting every ripple. (Reader, c'est moi.) I'll admit that it's not entirely impossible that at the tail end of 2021 I might have got a little carried away in my rakali quest. (For starters, I called it 'rakali quest'.) As soon as loosened lockdown restrictions allowed it I went for a three-hour dusk walk along my local stretch of the Yarra, from boathouse to boathouse across the river at the Fairfield Pipe Bridge next to Fairfield Boathouse, downstream to Studley Park Boathouse and back across the river at Kanes Bridge, then back home across Merri Creek to Clifton Hill. Of course the thing about a three-hour dusk walk is that dusk doesn't last for three hours, so most of it was actually done after dark, sweeping my spotlight across the river, hoping the rain didn't get too heavy, hoping I didn't slip. (It didn't, and I didn't.) I saw absolutely zero rakali. Still had a nice time, though.

A little later I was walking part of the same route, near the Pipe Bridge, when I did see a rakali: just a few seconds of it swimming towards the bank I was walking along before it disappeared. And of course now that I had an actual animal in an actual location my obsession only intensified. I went back to that spot, again and again; even took friends, including the friend who'd seen the rakali near Studley Park Boathouse. I saw the rakali – my rakali – one more time, a little way upstream under the Eastern Freeway; to this day I've never seen it again.

But I remembered another social media post I'd seen, a few weeks or a few months before – time moved differently in lockdown – about rakali in the Royal Botanic Gardens. A sunny day came along. I took the day off work. I used my allotted lockdown exercise time to ride my bike into the city, along the river, scanning the water the whole time. Riding along the pontoons on the river parallel to CityLink I saw the wake of something swimming disappear under the pontoon, and though I waited and watched it never emerged out the other side. When I'd seen the rakali under the Eastern it had similarly disappeared without a trace. Could this be …? What else could it be?!

I arrived at the gardens, locked my bike up at the gate and made straight for the ornamental lakes. These lakes used to be the Yarra, I'd read a few years prior, before the river had been straightened and this old river water had become isolated. I saw eels in the lakes once. I imagine an eel would make a good meal for a rakali.

I walked all around the lakes, scanning the water. To my frustration some small coves and inlets were difficult if not impossible to see into, such were the shapes of the shoreline and the positioning of garden beds. The Instagram post I'd seen had made a point of mentioning that new garden beds on the lakes had become a favourite haunt of the gardens' rakali: the animals were more easily able to climb up and down off them than they were the sheer stone sides of the original garden beds around the lakes. I paid particular attention to any of these new garden beds that I could see, but they were all disappointingly devoid of giant semi-aquatic rats. I sat down on the grass to have some cake, while fending off some overly familiar purple swamphens.

And then, right in front of me, a plop of water, and a white tail disappearing under the surface like a length of spaghetti being slurped into someone's mouth! I jumped to my feet and rushed to the edge of the lake, but the rakali was gone. I waited and waited for it to resurface, but it never did. I had a sighting though! In broad daylight! They were here! I finished my cake, picked up my binoculars and off I went.

I'd never previously paid much attention to the layout of the lakes in the gardens; I hadn't quite appreciated that it was a series of small and large lakes, interconnected by narrow waterways spanned by small bridges. Lots of nooks and crannies for a rakali to swim into; lots of places for it to disappear to nibble on an eel. I'd already walked around nearly the entire lake complex and got only a fleeting glance of a tail, and I was acutely aware of my time ticking down before the lockdown regulations said I had to be back inside – but I couldn't go home without properly seeing the rakali! Having had that fleeting glance was almost worse than having had no glance at all. (No it wasn't, it was awesome; I'm just being melodramatic.) I continued walking around the water, scanning every bank.

No matter the weather, no matter the season, no matter the time of day, there are always lots of people in the Botanic Gardens. Quite a lot of them have cameras. Quite a lot of those cameras are big chunky things with long lenses for taking photos of wildlife. But the gardens being such a busy, public space, there's not the tense silence of other areas where people go to take photos of elusive animals. Which is all a long way of saying that I saw some people taking photos of something, and I asked them what they'd seen. And it was the rakali.

'They're pretty tame here,' one of the people said. It's often like this: animals can get accustomed to people, if they're in places where people constantly are, and they become emboldened. The rakali was sitting in the sun on one of those new garden beds I'd been looking for, and it was only 10 metres away. The other people continued on and I settled in to watch, while it went about its business completely unfussed.

It dived into the water over and over again, slipping in like a spoon dipping into honey. It would swim around a bit, looking for whatever it could find, and a line of bubbles would sometimes rise to the surface; then it would reappear, sliding out of the water just as effortlessly and as smoothly and as quickly as it had gone in. Its dense fur was spiky wet; it was enormous, easily as big as the ducks it shared space with, and it was charming: fast and busy in that particular rodent way, never still, seemingly living in fast-forward, always up to something even if that something was just grooming. When it was on dry land it was hyperactive – but when it swam it was effortless. I don't know if you've ever thought about it but a rat shape is beautifully streamlined: sleek and tubular, long. It's strange to think of a rat as an apex predator but there wasn't a thing in those lakes that could have eaten that rakali. In his novel *Flames* Robbie Arnott positions a rakali as the god of its river, and when you see one in real life going about its business it makes perfect sense. Everyone should see a rakali at least once in their life – especially people who have a fear or loathing of rats. It'll turn them right around, I'm certain of it.

Anyway, my time came up and lockdown rules said I had to go back home, so I did, with a song in my heart and a grin on my face a mile wide. The sun was shining and I'd completed my rakali quest and I'd stayed out longer than I was legally allowed to because how could I tear myself away, and if somehow I got a fine then it would've been worth it because I'd seen the golden fur of Australia's biggest and best rodent, this continent's (and New Guinea's) own little big god of the waterways, swimming in ancient Yarra water.

THERE IS NO 'THEY'

Dogs were shooting like arrows down the grassy hill from the upper bend of the track straight into the river below. A gathering of people, swimmers and picnickers and joggers and dog walkers, were coming and going as if washed downstream by the river itself. When I'd contacted Danella Connors to arrange to meet for a conversation, and had suggested Deep Rock, she'd said, 'Perfect.'

The river was calm and gentle in the warm sunshine on the early January afternoon that we met, but only a couple of months earlier it had been higher than either of us had ever seen it before. Danella indicated the large flat-topped stones scattered about by the river for sitting on, and told me, '[After the flood] I shovelled dirt off them. Every one of them. I scrubbed every single one of them ... And someone said, "What are you scrubbing those rocks for?" and I said, "There's people that come here with their elderly parents, they park at

Dights Falls, and that is maybe the last or the second-last walk that that old person [will do] – they're coming down here not knowing [that], except that they want to get [outside] – and they'll sit on those rocks with their really frail old parents and have a little moment with them, and I just kept looking and thinking, "Imagine if someone did that and they get down here and there's nowhere for that person to sit." That was a lot of what went on for me, just my own personal response, going, "I want them to be able to sit and I want it to be nice and clean for them." It's so gorgeous though.'

I'd become aware of Danella when I was on Instagram and the algorithm suggested her account, Birrarung Clean Up, to me. When I met her I told her I'd only learned about it a week earlier. 'And it's been up two weeks,' she replied.

'I've lived nearby for twenty years,' she said, 'and I just quietly go about picking up rubbish and noticing things, doing something about it.' Spending time with her, even for just an hour that afternoon, it was clear that she's a thoughtful person – perhaps more thoughtful than most. She continued: 'And I just watched people's response to the recent floods, I just observed and noticed that some people couldn't wait to get back here and do exactly what they were doing before and there was no breathing space for what I call trauma. I felt like the river went through a really massive [trauma] … it was carrying a volume of water that it normally doesn't and everything was affected.' While the river was flooded precautionary tape was stretched across the broad path that runs past Deep Rock, and much of the path was under water – but even so, 'I watched how many people just were ready to come and get back on their jogging path … And then I had all these conversations with people and time and time again people kept saying, "[T]he river, the flood … the rubbish, I wonder what *they're* going to do about it?" And one day I thought, "I can't handle that any more, I can't handle standing with someone who is looking at rubbish within two arms' reach going, "When is someone else going to pick it up?" So I thought, "I'm going to clean up. I'm going to clean up the river and I'm going to talk to anyone that wants to listen."'

So, she told me, that's exactly what she started doing. 'I just brought a bag down and people would say, "What are you doing?" and I'd say, "Cleaning up the river" and they'd go, "What?" … Funny things would happen like I'd be bending down on the side of the path and … they'd look at me and see [my] grey hair and go, "Are you all right?" like I was having a heart attack or a stroke on the path, and I'd go, "Yeah, yeah, I'm fine … I'm picking up rubbish." "Oh!" I saw how many people's minds it threw just to see someone would pick rubbish up.

'And then I had things happen like people say[ing] to me … "You know what I just noticed in the last couple of days? I don't know what's going on around here but this path is really clean from the bridge up to here!" and I'd go, "Really?" I'd go, "Yeah I noticed a bit," and then they'd go, "No it's *really* clean!" and I'd say, "What do you think's going on?" [They'd say,] "I don't know, maybe they sent a crew down to clean it up." … And I'd say, "Maybe someone else cleaned it up." "Oh I don't think anyone would come down here to clean up the river!" I said, "Well you never know …"'

Eventually, Danella recounted, she'd give the game away: '[I'd say,] "Actually, there is someone cleaning it up …" and they'd go, "*Really?*" and I'd go, "Yeah," and they'd go, "Oh, do you know them?" "Yeah I know them quite well really." And then I'd go, "It's actually me."'

In the one-woman play Danella enacted for me based on her own experiences, this was the crux: 'So I'd say, "This is what I'm doing, I've got a bag with me, I'm just picking up the rubbish." … I just started saying to people, "Look, there is no 'they'. There's only us." Every day, they over there—' she waved her hand at a distant part of the riverbank '—they're just another bunch of us on a different side of the fence, or a different part of the grounds, or whatever. And the conversation has just taken off.'

With Danella organising the clean-ups, it didn't take long for other people to join in. 'Do you know Galatea Point?' she asked me. 'Do you know the big pile of rubbish [there]?' For a long time a floating patch of garbage has lurked on the river close to Galatea Point, a beautiful area of restored bushland near Studley Park Boathouse. 'That was one that was distressing me, it was really distressing me, so many people were saying, 'It's Parks, it's council, it's CIA,' everybody they could think of was going to come and clean it up, I don't know. And then I'd just end up saying to people, "I'm picking up rubbish and I want to clean up that patch on the river."' So she did.

'I just thought, "I'm sticking an Instagram page up, I'm going to put a call out," and I started. I went and talked to the Boathouse, they were happy to give us canoes and jackets with a waiver; I went to the golf club – I had to go to them a few times – they were happy to let us use the toilets, use the car park, and to give tea and coffee on the day ... I found [people would] go, "Oh are you from an organisation?" I'd say, "No, I'm just a woman who lives in Collingwood who's really concerned about what's happening on the river." And people had no comeback to that, because they couldn't say, "Oh can you get your sustainability officer to ring our sustainability officer?" or whatever, and so on the 29th [of December 2022] we had a crew of people that rocked up at six in the morning, rowed boats down to Galatea Point, lifted forty garbage bags of rubbish off that one little site.'

Danella's organisational efforts had also extended to getting a code from the local council to access the nearby recycling depot. Not everything they lifted out of the water was suitable for leaving at the depot, though: 'We pulled out a massive drum of acrylic acid and a massive bottle of chemical something-or-other – that got lifted off the river today, so nine days later. I rang, I was onto everybody I could find, and eventually – all of the EPA, Parks, everyone was decent but the process was very slow, and [Melbourne Water] finally sent a crew of contractors up the river yesterday ... and I was standing here this morning and they came down with the drum strapped on to their little flat-bottomed tinny.'

The Instagram page has helped: '[A]fter we did that clean-up [people contacted me] around the clock for like four days. Message after message after message after ... "When's it happening, can I help?" "I could do this, I could ..." People coming forward in a really organic, heartfelt way. And I've said to people, "Look it's up to *us*, all of us, we're all in the mess, we've all got to contribute." There is no "they", there's only us, so I don't want to deal with people being judgemental and blaming: "It's the ball throwers, it's the dog walkers ..."' That refusal to judge extends to her own clean-up efforts – recounting the clean-up at Galatea Point, Danella explained: 'It was a cross-section of ages, it was men and women; everybody worked really hard on the day, nobody complained. I just set it up and said, "You know what your needs are, you know how to keep yourself safe. Stick to your pace, you don't have to go at anyone else's. If you need a break, have a break. If you need something to eat, have something to eat. If you need a drink ..." I just wanted everyone to have space to feel like they were valued and like whatever contribution they made on that day was going to be really appreciated. And that if they needed to disappear for half an hour that was okay.'

I shared with Danella my own experience of growing up among rivers in Canberra, of feeling sometimes held at arm's length by Melbourne – but finding a way to embrace it through the river. 'I remember when I first came down to Victoria,' she said, 'someone told me about Dights Falls. I'd been living up in Cairns, and I came down – I remember seeing Dights Falls [which is only a metre or two high] and I burst into tears. And I sat there going, "So this is what people in Melbourne call a waterfall." I'm like "My goodness, I'm on another planet." And then over time I got to know the river a little bit more, a little bit more ... It's an incredibly wise being flowing through this country, really.'

In our conversation Danella noted wryly: '[The river's] 242 kilometres long, so if I had help it'd be great because I can't do that, I don't think I'll get that done in my lifetime.' For the time being, at least, her ambitions were more modest: 'I just feel like the way that this is happening there's a potential for this area at least, from say Dights round to Kanes Bridge ... maybe in time it could be a stretch of parkland that the community all participated in reinventing and recreating and taking care of, and what I'd love is if people just started going, "We take bags when we go walking, we pick up rubbish."'

Before meeting Danella, I'd joined a clean-up at Deep Rock organised by a mutual acquaintance. It was a Saturday shortly before Christmas 2022, and sun was glinting off the water. Bags and rubbish pickers were provided, and several people rowed kayaks down from above Dights Falls where there's a boat ramp into the broad, flat expanse of water at the confluence of the river and Merri Creek. I somehow found myself provided with one of two inflatable canoes – a single-seater, the other being a double – and set off uncertainly across the river from the oversized stepping stones at Deep Rock. Streamers of torn and faded plastic hung from every tree in sight along the river's edge and I made for one directly across from the steps. I was finally at the foot of the cliff I'd gazed lovingly at from the opposite bank so many times before.

I'd never been on the river before; *in* it, yes, a couple of times by then, but never *on* it. While I was in the canoe, a small puddle of water pooled at my feet from my awkward effort at launching, a Pacific black duck bobbed right up to me and looked me in the eyes, seeming to size me up, before swimming away. I've never been so close to one before, not even in a park where they have become habituated to humans. The iridescent panels in its wings flashed green, then blue, then purple as it changed course to swim past me.

I was wearing gloves and as I started to unpick the plastic from the tree I realised immediately that it wouldn't be as easy a job as I had imagined: the rubbish wasn't just draped over the tree, but wound tightly around the twigs and branches by the force and turbulence of the recent flood and then caked in mud. At one point a tiny spider ran out from under one of the coils of plastic, which it had made its home – a home I was now dismantling. Most of the rubbish I unpicked was anonymous plastic, bleached blank by sun and surging water, but I also – and at this point I was very glad I was wearing gloves – collected the remains of three condoms. 'Well I've found them too!' Danella assured me cheerfully when I told her this, a few weeks after the event.

During the clean-up I participated in, I spent the entire time on that one tree. I became fixated: there were so many trees draped in so much rubbish, but I was one person and I could at least do one tree. I'm not very good in a canoe but I managed to manoeuvre around the horizontal branches of the tree, wedging the canoe so that the river's current held me against a branch. At one point one of the other cleaners, one of the group who'd paddled up from Dights Falls, joined me, but she soon became bored or frustrated by the tree and moved on to the next one. For her Galatea Point clean-up Danella was on the bank – 'because I had to do phone call stuff and we had a cart on the bank that was loading up bags to go along the path and up the hill' – and she recalled, 'At one stage I looked out … and a tree was lying over the river with branches and the boats were poked in like this, and I looked out and I thought, "It's like watching bees on a banksia flower," it was almost just buzzing, [people] doing everything they could and they'd bring the next load of rubbish over and we'd haul it out of the kayaks.'

Danella cares deeply about the river – but unlike some who

care about it only insofar as they care about what they can get from it, she worries about the river, too, and about what it experiences. In naming her operation 'Birrarung Clean Up' she made a deliberate choice to use the river's original name, the one by which it's been known for thousands of years. In an Instagram post on 3 January 2023 Danella wrote:

> When I learned about how the river was named The Yarra, it didn't take long for me to consider and feel how that must have been for First Nations people, then and since ... The Birrarung is a living entity ... I wonder how she has felt not hearing her own name for so long? ... I've chosen to use her true name. It feels more loving and more respectful.

When I discussed with her my reasons for referring to the river by its colonial name in this book, she said: 'I get it, that's the river people have known. They haven't known the Birrarung. That's a whole new relationship, I think ... We're just trying to find our own truth within the circumstances. Now that I've really just thought, "Oh, this is the Birrarung," it's been overshadowed by this other thinking and this other name and this other way, so if I start to let myself slip in and know that I'm meeting the Birrarung, who am I meeting? What's this river about? What have I got to learn? What am I being given and taught and shown? Because I feel like I'm constantly, constantly, constantly being taught things when I'm down here.

'I love this place,' she said. 'I love this river, I love what it gives, and [after the floods at the end of 2022] I was hurting. I was actually really hurting. I'd be standing here going, "There's distress in the river, there's distress in the land." The river can't get all of that plastic out of the trees. And it had really taken a beating with the volume of water [that] was let out of reservoirs upstream. And I thought, "What happens in trauma? You've got to acknowledge the trauma and then there's got to be time to rest and recover."'

I asked her how she'd feel when the next flood dumped a whole lot more rubbish along the river. 'I don't know,' she replied. 'I don't know until I get there, but I'll be thinking, "Well I'm really glad we did last time." I'm realistic as well, I don't think that once we clean up it's over and done with. We're in an ongoing process of creating cultural change, and really getting to the space where people can see their own attitudes and go, "I can pick up three pieces of rubbish." Because I say to people, "Pick your timeframe. I do it every day; if you could do it once every six months, do it."'

GREY-HEADED FLYING FOX

In the first year of the pandemic a certain then-politician who's not worth naming kicked up a fuss about the colony of grey-headed flying foxes that were encamped in his electorate along the Yarra. He wanted to cull them because he thought – or claimed that he thought, honestly, who can tell – that they were a public health risk. Never mind that they'd been there for nearly twenty years without incident; never mind that they're a protected species and a threatened one, too; never mind that tens of thousands of giant fruit bats streaming out across the city at dusk on a warm night is one of the best wildlife spectacles it's possible to witness. As it turned out, half the colony was on the other side of the river and in the other major party's electorate so more fool him. Emphasis on 'fool'.

Anyway, that's enough about that, let's talk about bats! Melbourne happens to be rich in bats – not just in numbers but in species, too. There are up to sixteen different species of microbats in Melbourne, including in woodland along the Yarra – in fact, rivers and creeks are great places to spot them because they love gobbling up all the insects that hang around above the water. Microbats are the 'classic' bats, the Batman bats: the ones that live in caves (or drains, or crevices, or in old buildings, or sometimes just under loose bits of bark), the little flighty ones that use echolocation to catch insects. They're the 'blind as a bat' bats (though they're not blind at all, but that's another story).

Then there are the megabats. These are the ones you're more likely to see in real life (as opposed to in Batman movies) because, well, they're mega. They're very, very big. Well except for the ones that are quite little, but again, that's another story. The ones you see in Melbourne, flying out every night by their thousands as you catch the train home from a long day at work or ride off to see a friend in a park or what have you, they're megabats. They don't use echolocation because they don't eat insects. They eat nectar, and nectar tends to stay put so echolocation would be a bit of a waste of effort when it comes to finding it. Better to use adorable great big eyes and a cute snuffly nose. Am I a fan? Of course I'm a fan. Who wouldn't be? Nobody worth mentioning (again).

The megabats we get in Melbourne have terrible names,

really. A lot of animals have terrible names, true, but not so many are unfortunate enough to have *two* terrible names. On the one hand, these poor creatures are known as flying foxes. Granted, they *do* fly – and very well – but, look, they're not foxes. They don't even look very much like foxes. They're furry and they have a pointy face but do I really need to list all the different mammals that are also furry and have a pointy face? I mean *really?* They're not even carnivorous!

Which brings us to their other common name: fruit bats. Slightly better because, yes, they do include fruit in their diet – chewing it up and squeezing out the juice and spitting out the pulp – but but it's not their main deal. It's like if I was called a potato crisp primate and yes, fair cop, I absolutely *love* a potato crisp, but it's more of an 'I'll eat them if they're there' thing or, at best (worst?), an 'I haven't figured out anything to have for lunch and I'm walking past a shop' thing. They're not actually my main diet, and as much as I love them I wouldn't *want* potato crisps to be my main diet.

So, 'fruit bats'. Yeah, but also kinda ... nah. You see, the thing is that what they actually like to eat, what they're great at eating, is nectar. Flower juice. They'll happily eat fruit if it's around but if the eucalypts are flowering then there are a lot more flowers than there is fruit. It just makes sense.

This means that they're an important pollinator in Australian forests. I had a chat to Megan Davidson from Friends of Bats and Bushcare about the bats one afternoon in a suburban park (which was probably due to get visited by the bats that night) and she pointed out one good reason why there are so many flying foxes in the colony along the Yarra: 'If you've got a pollinator you need a lot of them, it's no good having half a dozen bees to pollinate – you need a shit-ton of bees. And same with flying foxes.'

And that's all fine if the bats are doing well – but the bats aren't doing so well. Remember how at the start of this chapter I said they were a threatened species? I asked Davidson about bat breeding and she explained: 'So females will breed probably in their third year. One pup a year, that's it. If they lose that pup or that pregnancy, that's it for another year. So very low reproductive rate, which is another reason why they're so vulnerable to big shocks: they can't replenish their population quickly.'

I promised I wouldn't bring him up again but another foolish thing our bloke who's no longer in parliament did was kick up a stink about a new sprinkler system that was built in the Yarra bat colony in 2022. The way some people carried on about it you'd think it was a ritzy health spa (which to be honest would be just fine as far as I'm concerned) but it's really just a series of long, tall pipes pumping river water up into the treetops when it gets really, really hot. And it's only getting hotter. 'Last summer in Europe Italy had a fifty-degree day, and in Canada they had a fifty-degree day,' Megan told me matter-of-factly. 'We will get a fifty-degree day in Melbourne. It's only a matter of time. And that's going to be ugly but we need to prepare for it. Do we want any of these animals to survive? If we do we have to be doing lots of revegetation, a sprinkler system obviously, every [flying fox] camp needs work. Because otherwise we could lose an entire species over one or two bad years.'

She explained what happens to a flying fox colony on an extremely hot day: 'They're homeotherms like us, they have to try to maintain their core body temperature within a narrow range. So if it's a bit hot they fan themselves, they pant; if it's a bit hotter still they start moving, they start getting out of the sun, trying to find a shady spot. It starts to get towards forty degrees, they're starting to really struggle to maintain their core temperature and ... they have a couple of behaviours. One's called clustering, which we don't understand, where they'll get into tight clusters in a tree and you'd think that that would not be a good strategy because of the heat within the cluster. Some people have got different theories about it, something about probably the surface area and the mass – I don't understand it. So there's clustering, but the most severe is what we call clumping, which is where they funnel down a tree, they'll end up covering the trunk and piling up on the ground and they just keep coming down and the living will come down on the dead. That's extreme – we call that colony collapse. So we can spray them and save quite a lot of them who are in that, but obviously once internal temperature reaches a certain level it will cause organ damage and failure in the same way as it would for humans.'

The colony hasn't always been in Melbourne, though the odd flying fox has been visiting the area for a very long time – maybe as long as the species has existed. 'You can find records going back pretty much as long as Europeans have been around,' Megan said. 'But of course the stories are all about how Mr So-and-so from So-and-so saw a bat and shot it, it's almost always that. But there are some really interesting early reports of bats in numbers in unusual places like up at Olinda. But evolutionary history runs into millions of years on this continent, so their range would shift with big changes in climate and habitat over time but they've been here for a very long time.' Nowadays grey-headed flying fox colonies are popping up everywhere: there's a colony in Canberra that definitely wasn't there when I was growing up in that city. There's one in Chiltern in north-eastern Victoria that wasn't there even when I first visited Chiltern in 2014. There's one in Bairnsdale. Et cetera.

I asked Megan if the colonies ever splinter. 'They do,' she said. 'They splintered when they were dispersed from the [Royal] Botanic Gardens in Melbourne – that's when the Geelong camp set up. And there's another little one at Lara. Werribee. There's another little one at Warburton in the redwood forest ... There's a little one at Bacchus Marsh, there's a little one at Colac, Warrnambool. So they need camps to be spaced 50 or 60 kays apart ... But [flying foxes] are constantly on the move, so the bat that is flying over my house today might be flying over someone's house in Sydney in a few weeks and someone else's house in Brisbane a few weeks after that.'

But back to that dispersal from the Botanic Gardens. It's a key part of the lore of the Yarra colony. It happened in 2003. Megan didn't have too many kind words to say about the Gardens or how they treated the colony in their midst. I asked her how the bats were resettled in Yarra Bend Park and she laughed: 'There was a fair bit of luck. You can't relocate flying foxes, you can only disperse them.' You can't tell them where to go, you can only make it so that they don't want to come back. 'So they dispersed them with noise and smoke and lights from the Botanic Gardens,' Megan explained, 'which is a big production. You've got to make it not a good place to be [for the bats], which means starting at about an hour before they would normally wake up and fly out and being there before dawn to make it not a nice place to come back to. So initially the bats just dispersed all over Melbourne, there were bats in every park in Melbourne – in Fawkner [Park], they were in the gardens at Treasury, they were everywhere ... so after some weeks ... there was some concern about how this was going. In one of the gardens somebody had gotten bitten because a branch had collapsed and he'd gone in to try and help the bats and of course gotten bitten so everyone went, "Oh shit, what are we going to do? Someone's going to get lyssavirus" ...

'That's one of the dangers of dispersing a flying fox colony: they may end up somewhere you like even less. The local hospital or the local nursing home. Anyway, so one day they were seen to congregate down on the river in Hawthorn, so that's when [the authorities] backed up and went, "Right, they're on the river." So then they had a nudging program, so every day they'd get downriver and make it a little unpleasant for them – not so that they'd fly off again, but just so that they'd start to move – so they did that for weeks and weeks ...

[T]he government had a list of acceptable sites: number one was Horseshoe Bend, number two was Yarra Bend. So when they got to [Yarra Bend] they went, "Nup, this is good. We're never going to get them across the freeway, this is good." And that's where they've stayed ever since.'

Though last time I visited the colony, at the start of 2023 just a few days before meeting Megan, I was surprised to find it not quite where I'd expected it to be. If you look up where to go see the Yarra's grey-headed flying fox colony every website will tell you to go to Bellbird Picnic Area in Kew, next to Yarra Boulevard. It's a lovely spot, and not just for the bats. There are viewing platforms, and information signs, and though you can walk for hundreds of metres in either direction on either side of the river and be walking beneath bats, the picnic ground is the epicentre.

Except when I took my girlfriend there because she'd never seen the colony before, it wasn't. I was a little embarrassed to find there was not a single bat at the picnic ground. Nor could any be seen on the far side of the river. In the end we walked for nearly ten minutes upstream before we found any.

'Well,' Megan explained, 'what happened last winter [in 2022] was, they've been at Yarra Bend Park for twenty years now – coming up twenty years – and they've always been between the Bellbird Picnic Area [and the Eastern Freeway] and then as numbers build they move up river. Last winter the overwintering animals – of which there's three thousand or so [that] usually spend the winter in Melbourne – they moved right up near the freeway. And when the animals were coming back as the weather got better, that's where they've stayed. So the whole core of the colony has shifted upriver.' I wasn't the first person to have noticed that they'd relocated. 'People say, "Why do you think they've moved up there?"' Megan recounted. 'Different microclimate? I think maybe, evolutionarily, because they have great site fidelity or camp fidelity, so they'll use the same camp possibly for hundreds of years ... if they every now and then move around their area they kind of make it sustainable because they're giving trees a rest ... That's the only thing I can think of that would explain it. In the hot weather they're still moving downriver, so on that really hot day they were down as far as the picnic area, but over the next couple of days they contracted back up.'

I got the feeling Megan could have talked about the bats all day. I could have, too, to be honest. She didn't mince her words but her face lit up every time she spoke about the bats. Part of the care she and her volunteers provide for the colony is rescuing orphaned and injured bats. At one point she showed me a photo on her phone: 'This is cute,' she said. 'This is my first two [baby bats] this year, they're so cute. They're big now. They're going to bat school on Saturday.'

Wait, there's a bat school?

'Bat school is where they go for their next step. So they're weaned and they're flying and they're getting hard to control, too, and they don't want to be in their enclosure; they want to be flying around the house which cannot happen because they crash into things. So they go up to Fly By Night bat clinic at Olinda, they've got a big aviary there, so all of the hand-reared orphans go there, and then they come back down to us at Yarra Bend in groups of about between thirty and sixty in age groups, so obviously the older ones go first.'

At Yarra Bend they're housed in a soft-release shelter where they can continue to build their strength and become accustomed to the presence of the other bats outside in the colony. Then, when they're ready, they're released.

I was lucky enough to witness the release of a graduate class of young bats back in March 2022, when Megan and I first met. I joined Megan and her colleagues just before dusk, down by the river, and she unlocked the gate and let me in behind the screened-off fence of the shelter. She and her team prepared trays of fruit, which they placed on the edge of the shelter, and then they opened the doors. Several bats were hanging upside down from the wire ceiling of the shelter and they started scampering with surprising speed and agility towards the fruit. Megan reminded one of the volunteers to look up before opening or closing the door because bats might be above her, and get accidentally trapped. The bats made for the fruit pieces and pouched them in their mouths, eyes and cheeks bulging with the joy of a quarter of an apple to chew on. Their ears and noses waggled like radar dishes. Then one of them suddenly took off with a whooshing sound, its wide, leathery wings beating the air just inches from our faces. It flew around the corner, outside but not quite ready just yet to commit to it, and hung from a low tree branch. Then it flew back and hung, instead, from the jumper of one of the volunteers. Megan gently coaxed it off, and it flew back to its branch as other bats emerged from the enclosure. It was controlled chaos, as any release of wild animals inevitably is. You can only control what you can control; the rest is up to the animals.

And I thought, as I watched these extraordinary animals flapping around at the level of my head, closer than I'd ever been to them before: *How can you not love bats?*

WE STARTED BECAUSE OF A FLEA

When you're watching a wild animal, you can bet that it's watching you, too. If you're calm and quiet and maybe sometimes a bit lucky the animal will decide you're okay, and will continue with its business, but at the start, from the moment it first notices you, there's a sizing-up period: it'll watch you, ears twitching or nose snuffling the air or just eyes fixed, patient, alert, waiting for you to make a move so that it can assess what kind of a situation you are.

When I met Sue McKinnon, having arranged via email to have a conversation with her about the impact of logging on the Yarra, I got the same feeling. A writer isn't automatically the most trustworthy person, especially one who pops up out of the blue and wants to pick your brain about a controversial subject, and I sensed for the first half of our conversation that Sue was sizing me up, watching, listening, figuring out what I was all about. And fair enough, too.

Which is not to say that the conversation was cold, not at all. 'I'm glad you read stuff,' she said, when I told her right at the start that I'd been reading about the various and ever-mounting controversies surrounding VicForests, Victoria's state-run logging agency. She asked me if I'd read *The Comfort of Water*, Maya Ward's book about her walk with friends from the mouth of the Yarra to its source – or at least as close as they could get to the source. With the catchment out of bounds, the group found themselves – after three weeks of walking – in a logging coupe.

Ward's book was published in 2011. Since then the logging of Victoria's forests has continued – sometimes illegally, and often despite the wishes of the people who live alongside the forests. Wherever there's a forest, there's a group of people devoted to protecting it. I asked Sue how her group, Kinglake Friends of the Forest, started.

'We started because of a flea,' she laughed. 'I haven't thought of this this way before,' she added, as she smiled and gazed back into her memory. 'We had Nillumbik Friends of [the] Great Forest and so we had quite a group anyway, and then ... [s]omeone offered to talk to us about a critically endangered flea that lives on the Leadbeater's possum. Only on the Leadbeater's possum ... this guy loves fleas and he offered to give us a talk, and you can never [turn] down a flea expert.' She laughed again. 'We thought, "We'll run that in Kinglake, it seems like a Kinglake thing to do," and we did and some people from Kinglake came and then we got to know them and we thought, "We'll start a Kinglake Friends of the Forest too," and we did.'

What started with a flea quickly scaled up, though. 'A little bit later Kinglake started to be logged in an area that we knew about,' Sue explained. 'It had been logged since the 2009 fires, but it was a bit further back into Kinglake and we just got upset in 2019 when logging started. So Kinglake Friends of the Forest – which we'd already started – just got very active ... there were people in there in the forest protecting, standing there in front of bulldozers, there was a rally outside.' Remembering the day, Sue reckoned there were 'nearly a hundred community members and there were probably ten forest officers there and the logging group and a truck came in and got turned around and – yeah, very ugly'. Then she corrected herself: 'Oh it wasn't actually ugly, it was very peaceful, there wasn't any violence, it was very well run, but there were a lot of very upset people and a lot of tears actually on that day, because ... the loggers continued to log, and we could see it from the carpark ... and if you've ever seen a tree that you're just admiring for its having been there for a hundred years and it's so solid, so big and so tall, and it just comes down so quickly, a hundred years is finished in a matter of five seconds. It's just very emotional. That was kind of the start of Kinglake Friends of the Forest – they came too close. They got us very cross and upset.' She laughed again.

I won't be so cheesy as to describe Kinglake Friends of the Forest as the flea that drove VicForests to distraction, but the thought did occur to me. I asked Sue what the state of logging was now, more than three years after that initial encounter.

'It's all very quiet,' she replied. 'So currently as of November the 11th [2022] all logging east of the Hume has stopped. So VicForests can log right over Victoria, [but] most of their logging was east of the Hume, which includes the areas that feed into the Yarra, and they were logging those areas very heavily ... So we kept on finding greater gliders in areas they were going to log and reporting [the gliders], getting a record of them and reporting them to the authorities and then pointing out the clause in the logging code that says they can't do this, and the government would say, "Oh no, you're reading the law wrong," and they'd just let [VicForests] log the coupe and log the tree that we'd found the glider in and log the home range that we'd found the glider in. So we went out – it's always gorgeous to be in the forest, it's a beautiful place, and to see gliders is also lovely, so we loved doing that but it was always [with] that hint of sadness because we only ever looked in logging coupes ...' Sue's eyes sparkled as she talked about the forest, and the animals in it. 'So that went on for years and then an opportunity came up to try that clause in court and we did, and we won. So we were right all along, they were wrong.'

Sue went into more detail on the court case: 'We tried two clauses, one was they can't basically destroy the population, and two [was] that they have to look before they log – and the judge agreed with us on both. So she said, well, one – if [VicForests] haven't been looking where they're logging, they haven't been following the law; and two – if they haven't been protecting what they find, they haven't been following the law. So they've been illegally logging, we know that, proved in court, and also she put in orders.' This last point is important: activists can argue all they like that forests are being logged illegally, but at the end of the day their words don't carry the same weight as a court order. 'If VicForests disobeyed those orders it's contempt of court and it's very, very serious,' Sue said. 'The CEO could get in criminal [trouble] ... So they've been very scared to disobey those orders, so they have to look properly for greater gliders and they have to protect them where they find them. So that's all they have to do.' It doesn't sound unreasonable. 'They could go out and log now if they looked properly before they logged and protected them when they found them,' Sue said. '[But] that's so far removed from what they normally do that they haven't been able to do it ... So they said, "Oh well, we'll reassess what we're going to do, we'll have to devise new methods, we'll have to research this, and in the meantime we won't log."'

It isn't only greater gliders that are protected under the orders; yellow-bellied gliders are protected, too. 'If there's a yellow-bellied glider seen they can't log any hollow-bearing trees and any yellow-bellied feed trees which have a mark on them. So there's a lot of restrictions that are triggered if a glider is seen,' Sue explained. Nor is it just wildlife that has to be protected: 'There was also a clause ... that as part of the protection of greater gliders they had to protect waterways, so that also stuffed them a bit too because there's waterways – there's drainage lines, temporary streams, streams all over the forest of course. Everywhere. Everywhere in the forest runs to somewhere and they're all waterways. So that was a problem for them too and they challenged that order and got it modified – they can now cross over a waterway but they can't log a waterway, which we're very happy about because that will protect some kinds of corridors for gliders.'

I told Sue about an experience I'd had in the middle of 2022. Acting on a tip-off, I'd gone spotlighting with a friend in forest near Healesville, in the Yarra Valley. Walking back and forth through one small area of the forest we saw at least a dozen greater gliders and one yellow-bellied glider. At first we'd thought nothing was going to happen. I'm accustomed to spotlighting for brushtail and ringtail possums, and Krefft's gliders, which emerge from their hollows as soon as it gets dark – about half an hour after sunset in Melbourne. But the greater gliders kept us waiting for much longer: an hour, even an hour and a half after sunset. The forest was well and truly dark by then, and cold, but the gliders were easy to see, not just because of their bright eye-shine under my red torch beam but also because of how huge and slow they were. Krefft's gliders dart about in the trees and are so small that they often disappear into the leaves almost as soon as you see them. Greater gliders, with their enormous, fluffy ears and long, pendulous tails, plod along the branches, taking their time, reaching for the eucalypt leaves that are their sole diet. They're ponderous, and often sit still on wide branches peering down at you. If you're lucky you might see one glide, which is a unique sight. Unlike other gliders, which have a membrane attached to their ankles and wrists, forming a rectangle when they launch, greater gliders have a membrane from ankles to elbows – so instead of stretching their arms out wide from their bodies, they stick their elbows out and tuck their front paws under their chin and glide into the night like a triangle. I saw two gliders do this together – several of the greater gliders my friend and I saw that night were paired up – and the image is fixed in my mind.

Relating this experience to Sue, I ventured the thought that in an old-growth eucalypt forest with sufficient large tree hollows the food supply for a population of greater gliders – eating, as they do, only leaves – must be effectively infinite.

'Nup,' she said straight away, shutting me down. I asked her, how come?

'Because eucalypt leaves are full of toxins,' she said, 'so they can only eat the ones that are coming from a tree that's growing consistently. I think it's the new shoots, but anyway it's trees that are growing in high-nutrient soil, because they need the leaves with lots and lots of nutrients in them because most of the energy they get from all of their eating is spent detoxifying the leaves.' This is one of the reasons why they're so vulnerable to logging: 'You cut down that [feed] tree [and] they've only got a limited number of trees they can go to within their home range ... they don't have the energy to go searching outside their home range ...'

Yellow-bellied gliders are different. '[They] don't have a home range, so they don't get their home range protected but instead they get these other protections because they travel,' Sue told me. She spoke fondly, grinning at just the thought of the yellow-bellied gliders. 'They scream around the forest, they're crazy ... Because they just [eat] sap so they're just full of sugar, they're just naughty little kids.'

I suggested to Sue that because VicForests is a statewide organisation, presumably the laws as affirmed by the recent court order would apply across Victoria. Sue explained that the court order only applied to the area covered by Kinglake Friends of the Forest: 'We cover the Yarra, areas that feed into the Yarra, Warburton, up to Baw Baw, and we cover the areas that feed into the Goulburn–Broken ... And that's all we have standing for.' But other groups have been fighting similar cases, particularly in East Gippsland, 'So East Gippsland was protected by these orders as well ... the logging in Gippsland and the north-east is so similar, they couldn't argue that it's any different. The gliders are the same.'

But the logging is not the same everywhere, so the court's findings can't be applied statewide. 'West of the Hume they log in a different way,' Sue explained. 'They log more for firewood. They don't normally take that wood back to the paper mill, so that wasn't covered at all [by the court order]. So what happened [after the court order] was all of a sudden VicForests got interested in trees that had fallen on the ground in the forest, Wombat State Forest, and started taking those to the paper mill. There was a lot of community upset about that. And now they're actually also doing that in the Dandenongs, they're about to do that in the Dandenong [Ranges] National Park as well – about to, that's all set up – but they haven't started yet I don't think, [and] they have done it in the Silvan Reservoir catchment. Again, people are upset about that because logs on the ground protect the ground – it's not the log that's the flammable part, it's the fine branches and they're the bits that VicForests leave anyway.'

As Sue was speaking, the clichéd criticism of environmental activists, repeated in a million tabloid headlines, came to me – that activists are all outsiders, imposing their city ideas on simple, hardworking country folk. I mentioned this to Sue, and observed that all the stories she'd told me were about activists fighting to protect the areas they live in. 'Exactly!' she said. '[VicForests] are the outsiders. Where's their office? In town? And ... at one stage after the fires there were loggers coming from East Gippsland down to Kinglake to log, they weren't locals. The crew that we were very upset with that first time when we came across the logging in Kinglake was from Romsey! They're not *really* locals.'

And the economics of logging native forests in Victoria don't even make sense: 'The thing is,' Sue said, 'all the logging has been for paper, so only 4 per cent ends up as timber. I mean people who particularly like blonde floorboards might think that's particularly valuable but it's meant that because this wood was available so cheap from the government, it's been subsidised, and VicForests lost $54 million last year – they've lost money the last five years. Since they started they've had over $100 million in grants and only give dividends back – $7 million. So they're losing money all the time, since 2004 when they started.'

To finish our conversation, I asked Sue what the worst- and best-case scenarios were for Victoria's forests, and for everything they sustain: worst case if logging continued as it has been, and best case if it stopped.

'There is a worst case,' Sue said. 'It would be devastating to be in a world that has consciously and knowingly made greater gliders extinct. To not be able to share the love of these beautiful, graceful, peaceful inhabitants of our forests. To know we did not allow them the right to live.' But, she explained, 'the Victorian mountain ash ecosystem is critically endangered ...

and that's because it needs to be twenty-five years old or so before it has seeds that are viable to regrow. And because when a fire goes through they'll probably all die; if it's a severe fire they'll probably all die and regerminate. And mountain ash can cope with that every couple of hundred years, but with logging added to that and then more fires it's possible we'll lose the entire mountain ash ecosystem ... [The] mountain ash ecosystem that's all around Maroondah [Reservoir] ... keeps that dam, keeps that water very, very clean. It's a wet forest, it has understorey of ferns and wattle and there's mosses that ... hold the water till [it's] needed or when they drain off to the riverland, the water reservoirs. So if we lose the mountain ash ecosystem – and that is quite likely, that's why it's been listed as critically endangered – what will replace it? I don't know ... about 50 per cent of the areas that they log in mountain ash don't regrow as proper mountain ash forest, they wouldn't pass the test of regrowth forest according to the logging laws ... There's been studies done. And I have certainly walked through an area that's been logged and it was just pure wattle. So what will that do to water?' Sue couldn't say for certain. It's worrying to think about.

As for the best-case scenario, Sue didn't give me an immediate answer. Instead she spoke more about logging, and at length about inappropriate fire regimes: 'There's absolutely no reason to do burning in forest unless it's right close to people's properties or coastal communities, and I say within 500 metres of communities. There's no reason to burn deep into the forest [as] they're doing, thousands and thousands and thousands of hectares ... it makes the forest more flammable, it burns with higher intensity, the shrub comes back after. There's been so many studies to say planned burning is not good, is not making the forest less flammable [or] burn with less intensity ... [And] logging makes the forest more flammable and burn with higher intensity, higher severity ... [Y]ou only have to look at a forest that's been regrowing ... they call it kindling forest because there's so many narrow stems and it's what you'd build your fire from. And the forest is always dry because those stems are very thirsty ... there's no moss, there's no tree ferns to put shade on the ground.' Ever ready with facts and figures, she told me: 'It's about twenty-eight years after a fire that the forest is less likely to burn again. Mountain ash, seven years after logging to thirty-five years after logging is highly flammable.' And then she got to the best-case scenario: 'So with the two of those combined [an end to logging and an end to burning the forest] we may end up having forests into the future [and if that microclimate was protected] maybe we'd go over that hiccup of whatever [the] maximum temperature of the world's going to be. I don't know if I'll be there,' she added finally, with a laugh.

And then gradually our conversation turned towards the Yarra. We talked about how good the water is in Melbourne, thanks to the Yarra and its forest catchments; she told me about her daughter's favourite places to swim in the middle Yarra. We talked about the wildlife of the river: the gliders, but also phascogales, and fish such as barred galaxias. And it became less of an interview and more just a chat, two people who love the river talking about the river. That in itself felt like something of a best-case scenario.

SUGAR (KREFFT'S) GLIDER

I guess the first thing to know about sugar gliders is that they're not all sugar gliders any more. Which is to say, they never were. Or at least not for a very long time – evolution being dynamic and ongoing and all that, common ancestors and divergence and what have you. Anyway, the point is, in 2019 a group of biologists did some digging around in the DNA and realised that the animals we'd been calling sugar gliders were in fact three different species: sugar gliders (the original), savanna gliders and Krefft's gliders. Krefft's is what we were left with down in south-eastern Australia, including Melbourne. So Krefft's they are, and will be, and I'll try to remember to call them that, though I'm good at forgetting because while I respect the science I'm not so keen on the name.

Anyway, that was all in the future when I was first told about sugar gliders in Melbourne (because they were still called sugar gliders back then). You might remember me mentioning it earlier: looking for platypuses, chatting to someone who was also looking for platypuses, them telling me that there were sugar gliders not 2 kilometres away. In fact there were probably sugar gliders 200 metres away, but I didn't know that then. Turns out they're more common in Melbourne than you might think.

But as far as I could tell they weren't in the tree that that person told me they were in. In fact the tree wasn't there, either. But there was one – more than one – that matched her description in another big bush park, just over the road, so that's where I started searching. It became an obsession, I'll freely admit – you might have noticed that's a bit of a theme in this book (as it is in my life, this book being at least 50 per cent me blabbering on about my life). So the gliders: it was painstaking, but also probably more fun than any obsession has a right to be.

I'd catch the bus up the freeway from my home – wonderful things, suburban buses, wildly underappreciated – and get off at a stop next to a servo; sometimes I'd go into the servo to get an ice-cream if it was a nice evening and I felt like it, then I'd walk up the road and duck through a stile in an old wooden fence and walk along the gravel track that led from the stile and before too long I'd be walking past old bush, old trees, old enough to have hollows in them. And I'd find a likely looking tree and I'd wait till dark, when the possums and other things started to emerge, and then I'd stake out that tree, watching, waiting, waving away mosquitoes, until I was sure that everything had emerged that was going to emerge, and if no gliders emerged then, or the next night, or two nights later, or three, I'd return and stake out a different tree. And it went on like that, for two or three months. There are a lot of trees. And then one night after finishing a stake-out I was walking along one of the paths through the bush park, spotlighting as I went, and there in a wattle I finally saw one: a sugar glider (Krefft's glider). Barely a glimpse. I took a terrible photo on my phone that convinced nobody except myself, but I knew. I knew what I'd seen, and more importantly I knew *where* I'd seen it. I had a solid data point now. I went back again, searched the same general area, saw another, started triangulating.

And then one amazing night, spotlighting at the right time and walking past the right tree, I saw them: tiny eyes, peering out of a tiny hollow. Way up high, on the underside of a branch. And out it came. Then another. And another. Because sugar gliders (Krefft's gliders) live in family groups, tamped down inside one little hollow, then they all emerge and go their separate ways for the night before all coming back home, or to a neighbouring home, because they keep a few hollows on the go.

I told a friend who was nearly as enthusiastic about it as me and we kept going back. We got familiar with the hollows, two or three of them in neighbouring trees. We came up with nicknames for them, so that we could easily direct each other to the hollow where all the action was happening on any given night – the best hollow was a vertical branch, dead wood in the heart of a living tree, which we called the Chimney. Night after night the glider family emerged one after another from the Chimney like Santa's elves on work experience, seeing how the big bloke does it. (There were plenty of brushtail possums around, too, much bigger and burlier than the gliders: Santa Clauses if we feel like stretching the simile to breaking and beyond.)

If you don't know where they're living, Krefft's gliders can be very hard to find. They're small, for starters: maybe half the size of a common ringtail possum. They're highly mobile: they can be gone from a tree before you even know it, and they can go a long way with only a few glides. They're surprisingly well camouflaged: their grey fur with a long, fuzzy black stripe down the back blends in brilliantly with the grey trunks and shedding bark of the eucalypts they like to spend time in. They spend most of their time in the canopy, feeding on blossoms.

And they're *fast*: even when they're not gliding, they flow up trunks and along branches like water. My friend and I rarely re-found the gliders after we'd watched them leave their hollows.

This agility is learned, though. Over the time that we watched the gliders we saw several young ones – tiny and tentative, seemingly left alone in the hollow by the adults, they peered out and gradually became bold enough to start exploring the immediate vicinity of the hollow entrance. And sometimes we saw these youngsters out foraging, clambering awkwardly through the trees, uncertain in their movements. One time one of these young gliders got so startled by the sound of my friend stepping on a stick that it nearly fell out of the tree. It's fascinating, and instructive, to see how the behaviours that we assume are innate are actually skills that need to be practised and perfected over time.

Spotlighting for gliders and other animals in that big bush park by the Yarra in the suburbs of Melbourne, I came to know the area fairly well. Walking a patch over and over again beds the knowledge of it down softly in your memory, pools it like water gathering in the hollow of a river-worn rock. I'd usually walk a loop, either side of a creek: away from the glider tree, over a bridge, down along one side of the creek, over another bridge, and back. One time crossing the second bridge I came face to face with a large and healthy wombat. And every time, by the time I got to that second bridge, I'd notice the same two interrelated things: the trees were younger and skinnier at that end of the walk, and the possums were completely absent.

At the other end of the track, large, old trees provided hollows in abundance, of all sizes to cater for all manner of wildlife – not just permanent homes for Krefft's gliders and brushtail and ringtail possums but also nesting sites for boobook owls and several species of parrot, large and small. The stark contrast between the abundant wildlife among the old trees and the near-absence of wildlife among the young trees, though the two zones were separated by only a few hundred metres, was eye-opening: on the Australian continent, wildlife needs tree hollows, and tree hollows only come with age.

If tree hollows aren't available, though, some animals are content to make do with artificial nest boxes. In February 2021 I was on French Island, south-east of Melbourne, helping to count waterbirds. Driving around the island that day I got chatting to the other people in the group, naturally, and naturally, given the demographic, we got chatting about wildlife: what we'd seen, where we'd seen it. I told one of my companions that day about the gliders in the large bush park by the Yarra and he told me that I should check out the park near where he lived, right next to a suburban train station and surrounded by houses. There were masses of gliders there, he told me.

So I did. It was on my local trainline, after all. It was an area I knew a little: I used to get off the train there and either ride my bike or catch a bus a couple of kilometres down the hill to another bush reserve on the banks of the Yarra where a pair of powerful owls could reliably be seen roosting. I'd even been in the very park that my French Island companion had told me about, at the other end of it, where for a few years in a row migratory and critically endangered swift parrots came to feed on eucalypt blossoms. When I returned to look for gliders I went with a friend and we kept an eye out in particular for nest boxes.

I can't remember how many gliders we saw that night, but we saw several, and they were easy to find. I've been back there many times since; it's less than twenty minutes on the train from my local station. I've seen gliders leaping from tree to tree while people jogged and walked their dogs past; I've heard gliders argue with each other and I've watched them chase each other out of trees – chase scenes better than anything any movie director ever dreamed up, with the victors enjoying the sweet nectar of flower blossoms and the banished making daring escapes by launching across open air high above the ground, gliding membranes stretched taut and long tails streaming behind. One memorable evening I staked out a den I'd found, a series of rotted cracks in the stump of a sawn-off branch two metres up a river red gum with townhouses only metres away, and I watched as glider after glider after glider emerged: before the mosquitoes from the nearby creek drove me away I counted eleven gliders from that one hollow. The same tree also houses families of ringtail and brushtail possums, and there's a family of tawny frogmouths nearby. The creek meanders its way through the park, goes underground for a while, re-emerges and then flows downhill into the Yarra. Even more than many natural phenomena, a river is a work of imagination as much as a geographic feature and to my mind the gliders in that park are as much a part of the greater Yarra as any animal that lives in the river's water. But if that sounds unconvincing to you, don't worry: there are plenty of riverside trees all up and down the Yarra that play host to gliders, too. I'm sure it's only a matter of time until I see a glider sail clear across the Yarra, from a tree on one bank to a tree on the opposite bank.

KEEP THE BIRRARUNG ALIVE

You may have noticed that there's an absence in this book. I certainly hope you have. If you've got to this point, you'll have read interviews with people who are swimming in the river, people who are revegetating the land surrounding the river, people who are helping manage the river's water. You'll have read stories of people who've come from interstate or overseas or just the next city over to live by and along the Yarra. What you haven't read so far though is the words of anyone whose people have lived along the river for so many tens of thousands of years that to even try to comprehend it bends my brain. The depth of knowledge accumulated like a riverbed over that time is unimaginable to me, as is the violence of the systematic gutting and degrading of that knowledge – the deliberate policy of cultural genocide.

Indigenous people have always lived along the Yarra for any meaningful and humanly useful definition of the term 'always'.

This book is not a book of history, though. I'm no historian; I don't have the skills or the patience. (History is a discipline and I'm undisciplined.) Besides, other people have already written histories of the Yarra, and done a much better job of it than I ever could.

This book is about now, and – like the Krefft's gliders in my local patch of Yarra Bend Park, which peer out of their hollow for an hour or more before finally deciding to venture out – this book is cautiously, hopefully about the future. Again and again, I asked the people I interviewed to imagine their best-case scenario for the future of the river. We all know the environment is in grave trouble; we all know, deep in our hearts, that we humans have a responsibility to correct the environmental degradation that modern Western civilisation has wrought. In Western capitalist culture we take and we take and we take, and it's well past time that we started giving back.

But if we're going to create a better future, we need to be able to imagine it. And to do that, we need good stories about the environment as well as bad ones: stories that don't greenwash the catastrophe that we're facing on this planet, but remind us to keep trying, to keep going, to keep striving. To fight for the best future we could possibly have. A deep future to accompany, and draw strength from, the deep past.

It may be difficult to truly appreciate the depth of tens of thousands of years of Indigenous history – it's a number I struggle to hold in the mind. But a future without Indigenous people and Indigenous knowledge at the centre of it isn't worth imagining.

Giving Indigenous people back their rightful space and access to Country is essential to realise this future, and it's a movement that is gaining steam. In December 2017 the Victorian Parliament passed the *Yarra River Protection (Wilip-gin Birrarung murron) Act*. Acknowledgement of Indigenous connection to the river is literally written into the legislation, right from the title; it was the first law in Australia to be co-written in an Indigenous language since European colonisation. According to the Victorian Department of Environment and Climate Action, the Woiwurrung phrase 'Wilip-gin Birrarung murron' means 'keep the Birrarung alive'. The Department's website explains further that 'Woiwurrung was used in recognition of the Traditional Owners' custodianship of the river and their unique connection to the lands through which the river flows.'

Among other requirements, the Act legislated the creation of a Birrarung Council to be the river's voice. This council advocates for the river, and advises the Victorian government in regards to it. Under the Act, 'at least two members of the council must be nominees of the Wurundjeri Tribe Land and Compensation Cultural Heritage Council'.

It's an important role, and it struck me as essential to learn more about it for this book, as well as the Act's other provisions – such as that 'Aboriginal cultural values, heritage, and knowledge of Yarra River land should be acknowledged, reflected, protected and promoted'. So I contacted the Wurundjeri Tribe Land and Compensation Cultural Heritage Council to learn more about their role on the Birrarung Council, and about how they felt about the future of the river. I hoped they would be as willing to talk to me as I was eager to talk to them.

Of all the myths about the creation of art, and about writing in particular, perhaps the most harmful is that of permissiveness: that in the service of art, anything goes. It goes hand in hand with another damaging myth: that the creation of art is a necessarily solitary pursuit, and that therefore every artwork is the creation of a single person. These myths are both held onto tightly, but they are just that – myths. This book may have my name on the cover, but its existence is the result of dozens of people working together: editors; designers; printers; interviewees. The bookseller or librarian who pushed this book into your hands is part of this collective effort. Without them, the book would be pulp.

And if the creation of a book is a collective effort, as the sole named author I must then consider the feelings of those people who have entrusted their thoughts and words and efforts to me. And similarly, I must be mindful of what stories I tell, and of whose stories I tell, and of whether my desire to tell a story is more important than somebody's right to privacy or self-determination (for me this is the simplest of all calculations: it never is, not for the kind of writing I do; I'm not here to play 'gotcha'). The more I think about it, the more obvious it becomes to me that First Nations sovereignty and self-determination are as essential to the wellbeing of this continent's future as they were to its past. Which brings me, in my circuitous way, to my reaching out to the Wurundjeri council.

When I contacted them, somebody got back to me and I talked to that person about why I was contacting them. I emailed them a list of questions that I was particularly interested in so that they might be able to find the best person within the organisation for me to talk to. And then months went by. I tentatively sent a couple of follow-up emails, and the person I'd spoken to got back to me with sincere apologies, advising that he was still trying to find someone suitable for me to talk to. My publisher and I discussed trying other avenues, which I pursued with similar delays. It felt facetious to write a book about the Yarra without including the voice of someone whose people have lived by the river for tens of thousands of years, who gave it its first and true name.

In mid-May 2023, still waiting for my contact to get back to me and amid a maelstrom of personal stresses and deadlines for the book, I attended a cultural walking tour along the river led by a Wurundjeri Elder. It was wonderful. The tour started at Abbotsford Convent and went to Dights Falls. During the tour, our guide mentioned casually to us that the Wurundjeri Tribe Land and Compensation Cultural Heritage Council has to turn down dozens of requests every day. That's not the number of requests that they receive, but the number that they have to turn down. Every day. Suddenly, my contact taking so long to get back to me made embarrassing sense.

I find myself – a white writer descended from a convict-turned-coloniser and living on colonised land whose first inhabitants have never ceded their sovereignty – increasingly wondering about the burden of asking Indigenous people for their input to my work. I've chosen to write about the environment of the Australian continent – it's what I like to write about, and I think it's what I write about best. But any writing of length about the environment on this continent as it is now that doesn't engage in an honest and meaningful way with Indigenous knowledge and the legacy of colonialism is not writing that's worth the effort.

What characterises colonialism more than taking and never giving back? And if that's the case, is my assumption that I can include Indigenous knowledge and words in my book – that I can ask an Indigenous person to become one of those uncredited collaborators and the request will not carry with it any burden – something akin to intellectual colonialism? I don't know. There's a chance that I'm being too sensitive, overthinking things. But I do feel like these are important questions for me as a writer to consider. I don't have the right to assume that all knowledge will be available to me whenever I need something to season my work. Though some people might suggest otherwise, this is not shame at the actions of my colonial forebears, or at the privilege that I as a relatively well-off white man of Anglo-Irish descent continue to benefit from within an ongoing colonial system. Shame is too often a paralysing emotion and, as such, I don't think it's useful when action against colonial oppression is urgently needed.

Rather, these thoughts come from a belief that white people

such as myself should be in a constant state of consideration and reflection regarding our place on this continent, which by default is also a place within a colonial framework. In the end, I never was able to ask my questions of anyone at the Wurundjeri council or the other contacts I pursued. After the walking tour I better understood why. But this has been my problem to solve, and mine alone. Just because I want to include something in the story I'm trying to tell, doesn't mean that I'm going to be able to. It's not such a bad lesson for a writer to learn.

During that walking tour along the river, the weather was grey but dry. Although the focus of the tour was the Indigenous significance of the river, the more recent colonial past was evident everywhere: from both the convent that we set off from to the remains of the old mill that we finished at; in the concrete roads soaring over the river; even in the plane trees dropping their leaves in the water, and the weeds choking the riverbanks. 'The cultural diversity and heritage of post-European settlement communities should be recognised and protected as a valued contribution to the identity, amenity and use of Yarra River land,' says the *Yarra River Protection (Wilip-gin Birrarung murron) Act*. We finished by the mouth of Merri Creek. If we'd walked a few minutes further we'd have arrived at Deep Rock, where the Yarra Yabbies swim.

'Community access to, and use and enjoyment of, Yarra River land should be protected and enhanced through the design and management of public open space for compatible multiple uses that optimise community benefit,' the Act says – though also that 'Public open space should be used for recreational and community purposes that are within the capacity of that space, in order to sustain natural processes and not diminish the potential of that open space to meet the long-term aspirations of the community.' The Act asks us to reflect not just on how we use the river, but also how we may misuse it – intentionally or otherwise. Wilip-gin Birrarung murron: keep the Birrarung alive.

It's now widely known that the name 'Yarra' sprang from a misunderstanding: when European colonists asked for the name of the river, they were given the word for flowing water. The river's real name is and always has been Birrarung: 'river of mists'. In the early mornings of June and July, when the leaves on the trees along the river hang damp and I ride my bike along the bank, the mist rises off the water, pale as manna gum bark. On some mornings the mist has been so thick that my sleeves have become white with it, the water vapour condensing on the wool of my jumper.

I've called the river 'Yarra' in this book because it's by that name that the river is so misunderstood; it's to that name that so many negative preconceptions have become attached. I thought that only once Melburnians had learned to see the Yarra afresh would we be ready to know it once again by its true name. I don't know if I was right or not. I love the river with my whole heart but perhaps I won't know it as I would like to until I learn to use its real name.

At the end of the walking tour along the river, as time on the afternoon's schedule ran down to zero and we turned to walk the brief length of river from Dights Falls back to Abbotsford Convent, I noticed unusual ripples and activity in the water of Merri Creek behind where we stood while our guide asked us if there were any further questions. I wanted to be absolutely sure that I was seeing what I thought I was seeing. I watched and waited until I saw the white tail-tip disappear beneath the water and the body of the animal resurface low in the creek, a v-shaped wake running out behind it before it dived again. Rakali! And only metres away! My heart beating fast, I stepped forward to get the guide's attention – I wasn't sure if this wildlife sighting was within the ambit of the tour, or whether I'd be wasting everyone else's time – but half a second before I could speak one of the other members of the party asked a question. Our guide, knowledge bursting out of him like the water of a flooding river, began to answer – and then elaborated, and elaborated, building a mosaic of story and history. The rest of the group listened, enraptured, while I kept turning to see if the rakali was still there, to see if there would still be an opportunity to share the sighting with the group. But as our guide started talking about his people's dispossession, the hardships they'd faced since colonisation, I realised that my divided attention was rude, and that I should be paying closer attention to what our guide told us. I stopped turning to watch the rakali, and I gave our guide my full focus. By the time he'd finished, the rakali had disappeared. We walked back to the convent, but I felt uncomfortable about my behaviour – about the disrespect that I might have unintentionally displayed to him. So as the group dispersed at the convent to go their separate ways, I took the opportunity to apologise to the guide, and to explain what I'd been looking at. 'You should have told us!' he said. And then he told me just the briefest snippet of information – barely even a fact, a fraction of a fact at best – about the animal's role in his people's culture.

But he didn't know that I was writing a book, and I didn't tell him, so that information will not appear here. It was shared with me, not explicitly in confidence but also not with the expectation that I'd share it with an unknown number of strangers. As tiny as it is, that information is not mine to tell. I thanked the guide, and I unlocked my bike, and I rode along the river back to my home, keeping an eye on the living water the whole time.

AZURE KINGFISHER

Around September or October, when the weather starts to warm up and the days are getting longer, the bush along the Yarra becomes loud with birdsong. All the birds that have hunkered down and stuck out the long, cold, dark months of midyear start puffing their chests out and throwing their heads back and singing their hearts out. And a host of birds that for one reason or another decided that flying up and down the east coast twice a year was a better option than putting up with Melbourne's grim midyear weather turn up and add their voices, too.

There are a lot of birds that are found all up and down the east coast but have decided that the Victorian coastline is just a bridge too far, enough of them for it to be a definite trend. Some of them, like blue-faced honeyeaters, are year-round residents in the state; others, like dollarbirds and rainbow bee-eaters, are seasonal migrants – they're perfectly happy to hang out on the coast at warmer latitudes, but in Victoria they're strictly inland, thank you very much. But there are just as many birds that are happy to come all the way to the south coast, including Melbourne, and many of them use rivers like highways. Including the Yarra.

I was walking in my local patch of bush in September 2022 – the same local patch of bush where I'd spent many happy nights visiting the powerful owl – when I noticed some unusual calls. A flock of yellow-faced honeyeaters, a conspicuously migratory species, were calling from the treetops with their characteristic signature phrase: 'Chip shop, chip shop, chip shop'. Scattered among them were smaller but no less noisy white-naped honeyeaters, with their own distinctive call like somebody slurping on a milkshake. For a few weeks, between these two species, the bush sounded more like a milkbar. A friend who went to check out the visitors thought he heard olive-backed orioles and sacred kingfishers, too: both species are seasonal visitors to Melbourne, both common along the Yarra in the warmer months of the year, both breed here before heading back north. Once in a reserve an hour up the river by bike I found, in a hawthorn tree that at other times of year had fed a roving flock of gang-gang cockatoos, an oriole's nest, now empty, and exposed among the newly naked branches. When in use it must have been hidden amid the hawthorn's dense leaves even though it was only at head height. Safe in the knowledge that the birds who'd once occupied it were now hundreds of kilometres away, I took a close look at it: it was a perfect cup of delicate twigs, finely woven, shaped by the warm bodies of the birds who'd built it. A couple of years later, not long after the yellow-faced and white-naped honeyeaters had visited my local patch, I heard from across the river at that same reserve the distinctive call of a shining bronze-cuckoo: another seasonal migrant, another river follower.

Azure kingfishers don't migrate along the river so much as commute along it. They're resident all year round, if you can find them, and they use the river twice over: as a highway and as a truckstop. They're so small and they zip up and down the Yarra so quickly that you almost never see them. There aren't many of them, at least not on the lower Yarra. Azure kingfishers are 'true' kingfishers, which means that they actually catch fish: they like to perch above the water and watch and wait, then when they see prey near the surface they dive down, beak first.

Azure kingfishers are the smallest of three species of kingfisher that you can find in Melbourne. The biggest, you'll probably have guessed, is the laughing kookaburra. Open a bird book if you have one, or just look up photos online, and compare the profile of a kookaburra to the profile of an azure kingfisher: if you ignore the colours and the size, it's pretty much the same bird. You've got a good chance of seeing kookaburras all up and down the Yarra, as long as there's bush. Kookaburras aren't 'true' kingfishers, and neither are sacred kingfishers, the third species. Which is a bit unfair because although they're not quite as brightly coloured as their azure cousins, sacred kingfishers are more impeccably kingfisher-like in appearance than kookaburras. Sacred kingfishers are another of those migratory species, and when they arrive in Melbourne you can hear them yapping all over: a four- or five-note call, repeated, 'Cheep cheep cheep cheep', which doesn't sound a million miles away from the calls that several other more common birds make from time to time, yet which is somehow distinctive in its insistency. Once they've navigated their way along the rivers and creeks of Melbourne they filter out into the surrounding bush, where they wait and pounce on anything that seems like it might make a decent meal: skinks, for instance. Poor skinks get picked on a lot: grey butcherbirds, smaller cousins of Australian magpies (though really Australian magpies are larger cousins of butcherbirds), love preying on skinks, and usually the first skinks that I see when the weather starts to warm up are in the beaks of grey butcherbirds. I've only seen two weasel skinks in Melbourne, and both of them have been dead: one was squashed on a mountain bike track, the other got pounced on by a butcherbird.

And if that's a bit too grim for you, consider this instead: one time, when that flock of yellow-faced honeyeaters was there, I walked to the top of my local patch and saw in a grove of profusely blossoming golden wattle a mixed flock of eastern spinebills, golden whistlers, grey shrike-thrushes, grey fantails, white-plumed honeyeaters, yellow-faced honeyeaters and silvereyes – the latter, with their distinctive plumage, having obviously migrated all the way across Bass Strait from Tasmania – busily feasting in the bloom. (I'm pretty sure birds don't eat wattle flowers, though, so they were probably visiting an apocalypse upon the many insects attracted to the wattle flowers – so maybe not such a beautiful scene after all, but oh well.)

A little further downstream from the inner-city bush where I saw all these birds, there's a reliable spot I've gone to many times over the years to see sacred kingfishers. Turning around and going up the other way you've got a chance of finding them all along the river in the warmer months of the year, and in Alphington and Kew you might find azure kingfishers, too: they've been seen in both places, from time to time. You might think that a bright blue and orange bird would be easy to spot among the more muted colours of the Australian bush but in truth they're so small and so patient, sitting on a perch and waiting for something to move in the water below, that they can be easily overlooked – unless they're calling while they're flying by, in which case they're gone almost before you can see them. There's one particular billabong along the Yarra that is a more reliable spot than most to see them – or at least to see one – but so far I haven't had any luck there. One time when I went there I mistimed my run: a bunch of people had seen the azure kingfisher, and taken photos of it, and then Melbourne experienced a momentous downpour of rain, so by the time I got there the billabong reserve was flooded. There were literally, without exaggeration, ducks swimming in the car park. Which was probably a bonanza for a tiny bird that likes to eat things it catches in the water but it made it even trickier than usual to find them – at least without a pair of waders or a canoe.

I've heard that azure kingfishers will take flight from a perch if someone approaches it from land but will sit unconcerned if someone approaches from the water. I don't know if that's true or not but I suppose it makes sense: azure kingfishers, I would imagine, don't have much to fear from the water. I guess there are fish that might have a go at them but they move so quickly and spend so little time actually *in* the water that it would take a pretty good effort to catch one.

I've seen azure kingfishers on the Yarra three times: twice while looking for platypuses, and once as far down as Dights Falls. There's a river red gum that hangs over the river where Merri Creek flows into the Yarra, just above the falls, and as I was crossing the footbridge there one day I heard a plop of water, turned around and saw a ring of ripples on the river and a small, vivid bird flying vertically back up to the lowest branch of the tree that spreads horizontally over the water. Now every time I cross that bridge I check the tree but so far I haven't seen the kingfisher again.

The first time I saw an azure kingfisher in Templestowe it was fishing at the mouth of Ruffey Creek. I wonder if the river, being so turbid, is hard for them to hunt in, and whether they favour creeks because the often clearer water provides a better chance of success for a visual hunter. Or perhaps it's just coincidence that two of the three kingfishers I've seen on the Yarra have been hunting on the confluence of a creek and the river (the third was flying fast and low along the river, busily on its way from point A to point B – that's the highway, not the truckstop).

I have a friend who likes animals as much as I do and in the warmest months of the year we both like to grumble about how everyone else sees snakes and we don't, and I have a theory, which is really more of a joke, that snakes only show themselves to people who don't want to see them and that they hide themselves from the people who love them. Certainly there are few occasions when you can go out and expect to see a particular type of wild animal – more often than not you're really just crossing your fingers and hoping. This is a pretty dubious hypothesis obviously but if nothing else it amuses me that I've only ever seen azure kingfishers when I haven't even been thinking about them, and whenever I've gone out with the intention or hope of finding one I've failed (it probably doesn't help that I'm the laziest birdwatcher in the world and can almost never be bothered getting up at the crack of dawn, when birds are most active).

But that's the story of nature, I suppose – and it's certainly the story of the Yarra: it gives you what it wants to give you, and you have to accept that with good grace, whether it's what you wanted or hoped for or not. And sometimes – more often than you might think – you get something amazing. It's just like people say, and yes it's a cliché but sometimes clichés are clichés for a reason: a river is life itself.

RESOURCES TO HELP THE YARRA

The Yarra Riverkeeper Association: yarrariver.org.au

Advocates for and stewards the Yarra, Birrarung River from source to sea so that it is healthy, protected and loved.

Friends of Warrandyte State Park: fowsp.org.au

Dedicated to the care of the State Park and the Warrandyte–Kinglake Nature Conservation Reserve.

Friends of Yarra Valley Parks: yvfriends.org.au

Dedicated to the restoration of the Yarra's parklands from Bulleen to Warrandyte.

Friends of Yarra Valley Flats: friendsofyarraflatspark.org.au

Dedicated to the restoration of the Yarra's parklands around Ivanhoe and Heidelberg.

Friends of the Helmeted Honeyeater: helmetedhoneyeater.org.au

Raising awareness about Victoria's critically endangered faunal emblem, the Helmeted Honeyeater.

Friends of the Leadbetter Possum: www.leadbeaters.org.au

Raising awareness about Victoria's critically endangered faunal emblem, the Leadbetter Possum.

Great Forest National Park: greatforestnationalpark.com.au

Dedicated to the creation of forest and protecting the Central Highlands of Victoria.

Westgate Biodiversity: westgatebiodiversity.org.au

Dedicated to transforming Westgate Park into a biodiverse, bush-like habitat for people in the city to enjoy.

Merri Creek Management Committee: www.mcmc.org.au

Dedicated to the preservation of natural and cultural heritage, and the ecologically sensitive restoration, development and maintenance of the Merri Creek and tributaries.

Darebin Creek Management Committee: dcmc.org.au

Dedicated to supporting a healthy and natural creek for thriving communities of plants and animals.

TRADITIONAL OWNERS OF THE BIRRARUNG

The Wurundjeri Woi-wurrung Cultural Heritage Aboriginal Corporation

The registered Aboriginal party covering Country through which the Birrarung flows. They bring their unique knowledge, connections and understanding of the river. Their interdependent relationship with the environment provides a blueprint for holistic management of the river and its lands.

The Bunurong Land Council Aboriginal Corporation

The registered Aboriginal party for Country downstream of the city. Bunurong Country covers the lower parts of the Yarra estuary out to Narrm (Port Phillip Bay).

RESPONSIBLE AUTHORITIES

Birrarung Council: birrarungcouncil.vic.gov.au/home

The council is a statutory body that provides independent advice to the Victorian Government on, and advocates for, protecting and improving the Yarra River.

Melbourne Water: melbournewater.com.au and 131 722 to report issues

Melbourne Water is responsible for managing and protecting Melbourne's major water resources.

Parks Victoria: www.parks.vic.gov.au and 131963

Parks Victoria manages 70,000 hectares of the parklands along the Yarra River, in particular Yarra Bend Park, Yarra Valley Parklands, Warrandyte State Park and Yarra Ranges National Park.

Environmental Protection Authority: www.epa.vic.gov.au and 1300 372 842 to report pollution

The EPA works to protect the environment and human health from the impacts of pollution and waste.

Department of Energy, Environment and Climate Action (DEECA): www.deeca.vic.gov.au

DEECA is the responsible Victorian government department and oversees more than 47,000 hectares of Yarra River land.